PERSPECTIVES ON WRITING
Series Editor, Mike Palmquist

PERSPECTIVES ON WRITING
Series Editor, Mike Palmquist

The Perspectives on Writing series addresses writing studies in a broad sense. Consistent with the wide ranging approaches characteristic of teaching and scholarship in writing across the curriculum, the series presents works that take divergent perspectives on working as a writer, teaching writing, administering writing programs, and studying writing in its various forms.

The WAC Clearinghouse and Parlor Press are collaborating so that these books will be widely available through free digital distribution and low-cost print editions. The publishers and the Series editor are teachers and researchers of writing, committed to the principle that knowledge should freely circulate. We see the opportunities that new technologies have for further democratizing knowledge. And we see that to share the power of writing is to share the means for all to articulate their needs, interest, and learning into the great experiment of literacy.

Existing Books in the Series

Charles Bazerman and David R. Russell, *Writing Selves/Writing Societies* (2003)
Gerald P. Delahunty and James Garvey, *The English Language: from Sound to Sense* (2010)
Charles Bazerman, Adair Bonini, and Débora Figueiredo (Eds.), *Genre in a Changing World* (2009)
David Franke, Alex Reid, and Anthony Di Renzo (Eds.), *Design Discourse: Composing and Revising Programs in Professional and Technical Writing* (2010)

DESIGN DISCOURSE: COMPOSING AND REVISING PROGRAMS IN PROFESSIONAL AND TECHNICAL WRITING

Edited by

David Franke
Alex Reid
Anthony Di Renzo

The WAC Clearinghouse
wac.colostate.edu
Fort Collins, Colorado

Parlor Press
www.parlorpress.com
Anderson, South Carolina

The WAC Clearinghouse, Fort Collins, Colorado 80523-1052
Parlor Press, 3015 Brackenberry Drive, Anderson, South Carolina 29621

© 2010 David Franke, Alex Reid, and Anthony Di Renzo. This work is licensed under a Creative Commons Attribution-Noncommercial-No Derivative Works 3.0 United States License.

Library of Congress Cataloging-in-Publication

Design discourse : composing and revising programs in professional and technical writing / edited by David Franke, Alex Reid, Anthony DiRenzo.
   p. cm.
  Includes bibliographical references.
  ISBN 978-1-60235-165-3 (pbk. : alk. paper) -- ISBN 978-1-60235-166-0 (hardcover : alk. paper) -- ISBN 978-1-60235-167-7 (adobe ebook : alk. paper)
   1. English language--Rhetoric--Study and teaching (Higher)--United States. 2. Academic writing--Study and teaching (Higher)--United States. 3. Technical writing--Study and teaching (Higher)--United States. 4. Writing centers--Administration. I. Franke, David, 1960- II. Reid, Alex, 1969- III. DiRenzo, Anthony, 1960-
  PE1405.U6D47 2010
  808'.0420711--dc22
                     2010001091

Copyeditor: Annabelle Bertram
Designer: David Doran
Series Editor: Mike Palmquist

The WAC Clearinghouse supports teachers of writing across the disciplines. Hosted by Colorado State University, it brings together scholarly journals and book series as well as resources for teachers who use writing in their courses. This book is available in digital format for free download at http://wac.colostate.edu.

Parlor Press, LLC is an independent publisher of scholarly and trade titles in print and multimedia formats. This book is available in paperback, cloth, and Adobe eBook formats from Parlor Press on the World Wide Web at http://www.parlorpress.com. For submission information or to find out about Parlor Press publications, write to Parlor Press, 3015 Brackenberry Drive, Anderson, South Carolina 29621, or e-mail editor@parlorpress.com.

This volume is dedicated to all those who are delighted by the study, teaching, and practice of writing.

Contents

Preface   ix

Composing   3

1 The Great Instauration: Restoring Professional and Technical Writing to the Humanities   5
Anthony Di Renzo

2 Starts, False Starts, and Getting Started: (Mis)understanding the Naming of a Professional Writing Minor   19
Michael Knievel, Kelly Belanger, Colin Keeney, Julianne Couch, and Christine Stebbins

3 Composing a Proposal for a Professional / Technical Writing Program   41
W. Gary Griswold

4 Disciplinary Identities: Professional Writing, Rhetorical Studies, and Rethinking "English"   63
Brent Henze, Wendy Sharer, and Janice Tovey

Revising   87

5 Smart Growth of Professional Writing Programs: Controlling Sprawl in Departmental Landscapes   89
Diana Ashe and Colleen A. Reilly

6 Curriculum, Genre and Resistance: Revising Identity in a Professional Writing Community   113
David Franke

7 Composing and Revising the Professional Writing Program at Ohio Northern University: A Case Study   131
Jonathan Pitts

Minors, Certificates, Engineering — 151

8 Certificate Programs in Technical Writing: Through Sophistic Eyes — 153
Jim Nugent

9 Shippensburg University's Technical / Professional Communications Minor: A Multidisciplinary Approach — 171
Carla Kungl and S. Dev Hathaway

10 Reinventing Audience through Distance — 189
Jude Edminster and Andrew Mara

11 Introducing a Technical Writing Communication Course into a Canadian School of Engineering — 203
Anne Parker

12 English and Engineering, Pedagogy and Politics — 219
Brian D. Ballentine

Futures — 241

13 The Third Way: PTW and the Liberal Arts in the New Knowledge Society — 243
Anthony Di Renzo

14 The Write Brain: Professional Writing in the Post-Knowledge Economy — 254
Alex Reid

Post-Scripts by Veteran Program Designers — 275

15 A *Techné* for Citizens: Service-Learning, Conversation, and Community — 277
James Dubinsky

16 Models of Professional Writing / Technical Writing Administration: Reflections of a Serial Administrator at Syracuse University — 297
Carol Lipson

Biographical Notes — 317

Preface

David Franke

This book grew out of the challenges of starting and sustaining a Professional and Technical Writing program at the state college where Alex Reid and I were hired (nearby, co-editor Anthony Di Renzo began his program at Ithaca College in New York a few years before us). We found ourselves building our program at the intersection of several academic and semi-academic discourses—rhetoric, English, new media, business, publishing, composition and others. We had plenty of theory from these fields and personal experience as students, teachers, writers, and freelancers. Yet as we established our identity as a major, we found that our interactions with other departments (especially English), our entanglement with the long-standing academic tensions between "liberal" and "vocational" education, the demands of staying abreast of new technology, the way our resources and students were distributed across many disciplines—all these pressures and others combined in unexpected ways, presenting us with a bit of a paradox in that we were compelled to make sense of the whole while we struggled with the day-to-day work of running a new program; simultaneously, most day-to-day decisions depended on a sense of our whole—our mission, rhythms, audiences, and strengths. Seen from a purely analytical perspective, what we were trying to do seemed impossible.

But of course it wasn't impossible. Our experience beginning a PTW program at the State University of New York at Cortland was typical in many ways. The undergraduate program we were hired to bring to fruition, like many others, was simply hard to define, lacking a deep sense of tradition that English and even rhetoric programs often enjoy. Our program was defined more by what it was not than what it was: not literature, not journalism, not composition. Despite this, the program grew, in part because we were able to invent an attractive curriculum, and our success introduced a new problem in that we were quickly understaffed: we had only three Professional and Technical Writing faculty in an English department of 50-odd full-time and part-time faculty. The demands on the three of us, all in new jobs, were sometimes intimidating. Actually, they were often overwhelming, as several authors in this volume have also experienced in their own schools. In front, we met the challenge of teaching new classes. At our back was an avalanche of paperwork. Struggling to keep moving forward, we found ourselves grasping for information and models. Like any academic in a new situation, we depended on our research skills first, and started reading.[1] The WPA (Writing Program Administrator) listerv (http://lists.asu.edu/archives/

wpa-l.html) gave us valuable clues to how writing programs run on a day-to-day basis, though its focus is of course more on Freshman English. National conferences, especially ATTW (Association of Teachers of Technical Writing) and CPTSC (Council on Programs in Technical and Scientific Communication), provided invaluable information about internships, key courses, recent theory—and at these conferences we found something the readings did not provide: warm, anecdotal, human stories. I sought first-person narrative accounts that presented the PTW administrator's logic and commitments, a constructive, sustained, intelligent set of discussions in relation to which we could shape our own history. To complete and understand our own program, we needed reflective stories that demonstrated and reflected on the process of making key, high-stakes decisions in the unfamiliar situation of running a professional writing program.

This narrative gap is what prompted my colleague Alex Reid and me to put out a call for papers that would, we hoped, assemble a community of narratives. Alex and I asked that PTW curriculum designers discuss how they composed and revised their PTW sites. We emphasized that we were looking for case studies in first person that revealed how designers made sense of and organized their particular location—in other words, how they historicized their work. Their stories would reveal the praxis of those in PTW programs working simultaneously as both teachers and administrators, often from the margins of English, Engineering, Composition/Rhetoric, and on the line between the liberal arts and professional schools. The focus was not to be pedagogical, but architectural, with an emphasis on design problems.

In its final form, each of the essays was to examine the complexities of developing, sustaining, or simply proposing non-literature curricula, from entire programs to individual classes. The authors were generally new assistant professors when these essays were written, and their contributions reflect an acute sensitivity to the practical contexts within which they worked—the political, historical, and financial realities—as well as a sense of vitality, a sense that something untested and unique could emerge and succeed at their respective locations. In the best pragmatic tradition, these essays explain how to both picture and perform a task, in this case the task of developing communities and curricula in PTW, with the belief that other designers might benefit from their narratives.

We experimented in this volume. Our always-supportive publisher Mike Palmquist encouraged us to go ahead with a form of peer review that helped us make the entire process as useful as possible to the authors and you, the book's audience. After outside readers gave the thumbs up to the book proposal, we solicited the essays. Alex Reid and I wrote responses to each essay we accepted and mailed our comments back to the author. Simultaneously, each essay was mailed to another contributor in the book for further response and comments. The results were strongly positive. Invested in the volume, peers generally commented critically and generously on one another's work and appreciated the additional feedback they received while revising. Doing so also helped contributors minimize overlap with other essays and gain a better picture of the volume as a whole. Conscious that many of our contributors are new to the field, we also invited several well-known figures in the field to read a grouping of essays and write "Post-Script" pieces based on their experience as program designers. Michael Dubinsky and Carol Lipson, experienced members of the field, graciously agreed to reflect on their careers in a way that gives context to the essays collected here.

Many of the articles collected here address what Robert Connors calls the "two-culture split" between the art and science of writing. That is, many of us struggle with practical answers to a question asked in various ways: are we to encourage insight or technique, liberal or vocational education, good citizens or good workers? This question is of course addressed by our theory, but has to be confronted also in even the most bureaucratic decisions about program requirements, a semester's course offerings, or even class sizes. This tension is also present every time a PTW faculty member sits down to write for publication. What balance does one provide for the reader between theoretical speculation and practical orientation? To put it another way, when we write for our colleagues in PTW, are we to provide interesting questions or interesting answers, the problematics of a course of inquiry or the results of a course of action?

The chapters here provide both, taking a stance that bridges the two cultures and often explicitly addresses the tensions between them. Faculty under the gun to organize a program do not have the luxury of waiting for the conclusion of big-picture arguments about the history, nature, and status of the field; likewise, short-term best-guess decisions won't sustain a program for very many semesters. Bringing together problem posing and problem solving is exactly what a program designer must do in order to begin and sustain his or her PTW program. This both/and thinking has direct application to the students' learning. The PTW programs here refuse to choose between teaching students to reflect or teaching them the skills to "succeed" – with "success" a term that teachers tend to think about even more critically than their students.

The 16 essays of *Design Discourse* are arranged in five sections. The first four chapters are grouped together under the heading of "Composing." Anthony Di Renzo's "The Great Instauration" addresses the practical and rhetorical challenges of setting up a PTW program in the humanities, addressing the chronic tension between liberal and practical arts. Drawing from Francis Bacon's *Advancement of Learning* in the opening essay, Di Renzo provides a theoretical and ethical framework in which "technical" subjects can serve as sites for the development and improvement of "social good." Di Renzo (like Bacon) appreciates the practical uses of knowledge, and eloquently turns Bacon's insights to pragmatic advice for those facing the challenge of beginning and beyond. Turning then to the concerns at a specific site, collaboratively written "Starts, False Starts, and Getting Started: (Mis)understanding the Naming of a Professional Writing Minor" (Michael Knievel, Kelly Belanger, Colin Keeney, Julianne Couch, and Christine Stebbins) historicizes the process of naming their minor as it unfolds at their particular institution over several decades. By tracing the various implications of their program's name, they present a nuanced study of how various stakeholders choose to interpret—and misinterpret—their program. They present the process of naming as an inquiry, guided by a set of ethical and practical questions, into their identity and audience: "are these expectations [raised by the program's name] at odds with each other? Which expectations can realistically be met given resources like faculty, funding, and goodwill?"

Two other articles in this first section discuss the process of designing in PTW in the face of serious challenges. As W. Gary Griswold puts it in "Composing a Proposal for a Professional / Technical Writing Program," writing the RFP (Request For Proposals or grant) for his program was a matter of "one week and five pages." A case study of the under-represented (and over-feared) process of submitting a grant application, Griswold's essay includes the original request for proposals and his response.

Completing this section, Brent Henze, Wendy Sharer and Janice Tovey's piece on "Disciplinary Identities: Professional Writing, Rhetorical Studies, and Rethinking 'English'" narrates their attempt to establish their proposed program in Rhetorical Studies and Professional Writing. The proposal itself was not well received. As they put it, they had inadvertently "thrown open the floodgates of disagreement about what a degree in 'English' means." Their candid narrative examines with equanimity not only the choices they made, but also what they might have done differently, making it useful to program designers who similarly have to traverse disputed academic territory.

"Revising," the second section of *Design Discourse* presents strategies for sustaining PTW programs. In "Smart Growth of Professional Writing Programs: Controlling Sprawl in Departmental Landscapes," Diana Ashe & Colleen A. Reilly develop an extended metaphor that draws on "systems thinking" from ecotheory and "smart growth" from city planning, using these schools of thought to guide their program's development. Their model promotes interdependence, change, and diversification as key principles that shape "sustainable and resilient programs." Presenting their attempt to strike a balance between specialization or succumbing to "the academic equivalent of urban sprawl," Ashe and Reilly's essay shows how a program can be both dynamic and principled as it develops an identity over time and in concert with various academic communities. My own essay studies change in our undergraduate PTW program in a small New York college. I draw from genre theory, which argues that established types of written texts, though they may appear "frozen" or inert, are in fact powerful and dynamic forces shaping a community. Yet I began the program with a fairly naïve understanding of how the curriculum-as-genre, as a published document, would function. I describe learning to work with that curriculum as an "enabling constraint," one that pushed us to evolve while also restraining our growth. Change is also the theme of Jonathan Pitts' "Composing and Revising the Professional Writing Program at Ohio Northern University: A Case Study. Charged with developing, sustaining, and creating coherence for his nascent major, Pitts shows how he deliberately planned for change without sacrificing coherence. His chapter includes the specific course offerings in his program and a vivid narrative of his experiences; it concludes with snapshot essays of several graduates from his program. In "Foundations for Teaching Technical Writing," Sherry Burgus Little explains that the "design and development" of certificate programs "crystallizes" the pervasive and long-standing debate over the ends of education (283). They inevitably raise questions about what sorts of knowledge is essential for students to do their work as PTW professionals.

The chapters in the third section of this book, "Minors, Certificates, Engineering," certainly confirm Little's insight. Though smaller than four-year undergraduate programs, these more concentrated sites introduce significant arguments to this volume, posing special problems for the program administrator. First in this section, Jim Nugent's essay "Certificate Programs in Technical Writing: Through Sophistic Eyes," the result of a survey of 62 certificate-granting sites, finds contemporary programs value "situated and contingent" knowledge that is both flexible, reflective, and socially engaged. Carla Kungl and S. Dev Hathaway present an adroit response to the pressure to professionalize in "Shippensburg University's Technical/Professional Communications Minor: A Multidisciplinary Approach." Recognizing the pressures on academic institutions to develop a "practical" writing degree, but lacking the resources or students to sustain a full-fledged program, they show how an interdisciplinary minor can gain a foothold. Their essay reveals how they juggle competing educational goals in their college, creating a "career-enhancing program for students while maintaining a meaningful liberal arts backdrop." Similarly, Jude Edminster and Andrew Mara in "Reinventing Audience through Distance" discuss the development of a program tailored to their situation, one with a large number of international students yet lacking local high-technology jobs. Their creative solution is to create a graduate certificate program that meshes with the graduate programs in Scientific and Technical Communication at Bowling Green State University. Rather than trying to prepare students for every specific technical task, these faculty teach their students to make decisions situationally. They draw from Thomas Kent and post-colonialist theory to articulate their approach, one in which students learn to "participate in meaning-making and to recognize their role in meaning-making."

The relationship between the humanities and the sciences is developed in Anne Parker's reflective essay, "Introducing a Technical Communication Course Into a Canadian School of Engineering: A Case Study of the Professional and Academic Contexts." There, she discusses developing a coherent and persuasive model for teaching writing that draws on the habits of thought internalized by engineering. Holding a position on the faculty in the Engineering school, she presents working as an "insider" to effect change there. Her chapter tacitly traces strategies for dealing with a complex and gendered institutional context. She also gives a helpful and detailed discussion of how to keep various elements of her course vital and interactive: her team, the collaborative process, and product. Also concerned with Engineering, Michael Ballentine of Case Western University shows us a successful approach for developing a writing pedagogy for engineers at his university. Dealing both with the graduate practicum course and the particular course for engineers that it prepares teachers for (over 350 students take it each year!), his "English and Engineering, Pedagogy and Politics" discusses the political and practical negotiations necessary to embed successfully an engineering program into an English department.

The penultimate section of the book, "Futures," is composed of two forward-thinking essays: "The Third Way: PTW and the Liberal Arts in the New Knowledge Society" by Anthony Di Renzo and "The Write Brain: Professional Writing in the Post-Knowledge Economy" by Alex Reid. Di Renzo's essay argues that PTW programs are a much-needed bridge for educational institutions torn between traditional liberal arts educational values and new pre-professional imperatives. PTW can provide an urgently needed social service by graduating rhetors with the know-how and eloquence to communicate between the various professions and disciplines, adept at responding to the demands of the new knowledge economy. Di Renzo's essay is essentially promoting a new image of what an "educated person" might look like, free of an affected disdain for worldly affairs or for intellectual play, and he argues persuasively that PTW programs are an apt site in which to begin education's "third way."

Likewise, Alex Reid's piece entitled "The Write Brain: Professional Writing in a Post-Knowledge Economy" confirms the centrality of technology for all PTW programs, placing it at the intersection of human and technical concerns. That is, Reid advocates for developing technical educational programs that draw from a vast range of intellectual and creative skills. He argues that several influences compel PTW programs to re-think their programs: the "knowledge economy" that has gone "offshore"; the consequent need for writers with rhetorical and critical skills; the rise of new Web 2.0 technologies which demand we teach students how to think "in" new media; the linked demands that Web 2.0 puts on us as faculty to teach and use such media to build knowledge webs and the like (Reid mentions wikis, blogs, and podcasts along with del.icio.us and flickr.com). His is not a repudiation of the humanistic, rhetorical tradition, but a reinscription of it (or "remediation" as Jay David Bolter might have it), accomplished in new media. Reid gives us a conceptual and pragmatic sketch of how these sea changes can and will affect our working lives in PTW programs.

Finally, in "Post Scripts" we have reflections from two experienced program designers, Carol Lipson of Syracuse University and Jim Dubinsky of Virginia Polytechnic Institute and State University. Dubinsky's "A Techné for Citizens: Service-Learning, Conversation, and Community" reflects on the decade-long process of creating an undergraduate PTW curriculum that is both practical and reflective, rewarding not only for the student but also for the student's community. He lays out the choices, both theoretical and practical, of designing a program that supports constructive civic action. The goal here is setting up students who can work with others on common problems, a harmony he likens to a form of reverence. Developing detailed and workable solutions to common problems is both a humanistic and technical commitment in Dubinsky's program, articulated clearly in this helpful reflective essay. Whereas Jim Dubinsky's essay addresses the process of getting up to interstate speed, Carol Lipson's reflective essay "Models of Professional Writing/Technical Writing Administration: Reflections of a Serial Administrator at Syracuse University" traces her journey through several different incarnations of professional and technical writing, stretching nearly three decades, at Syracuse University in New York. Her experience clearly contrasts two paradigms. In the first, program leaders are segregated and pursue somewhat independent paths in a clearly defined hierarchy; in the second, the leaders of various initiatives are (ideally) peers who share a complex and intertwined set of partially overlapping agendas. Hierarchy is less explicit, if not absent. Lipson's essay is candid about the complex institutional and administrative challenges that faced her as a PTW program designer, and gives a trajectory of her academic career which new PTW leaders will find useful and interesting.

We believe new program designers engaged in the process of sowing and cultivating their own programs will find in this volume's narratives something parallel to a reflective community, one that can help them develop their own program's identity, habits, and goals. We believe PTW programs can and do function at the intersection of the practical and the abstract, the human and the technical. It is our hope that the essays reveal these binaries working dialectically for the better.

NOTES

[1] We found the following texts particularly helpful: Katherine Adams' *A History of Professional Writing Instruction in American Colleges: Years of Acceptance, Growth, and Doubt* (Southern Methodist U.P., 1993); Teresa C. Kynell and Michael Moran's collection *Three Keys to the Past: The History of Technical Communication* (ATTW, 1999); *New Essays in Technical and Scientific Communication: Research, Theory, and Practice,* edited by Paul Anderson, R. John Brockman, and Carolyn Miller (Baywood, 1983); Katherine Staples and Cezar Ornatowski's *Foundations for Teaching Technical Communication: Theory, Practice, and Program Design* (ATTW, 1998); *Coming of Age: The Advanced Writing Curriculum,* edited by Linda K. Shamoon, Rebecca Moore Howard, Sandra Jamieson and Robert A. Schwegler (Boynton/Cook Heinemann, 2000).

WORKS CITED

Bolter, Jay David, *Writing Space: Computers, Hypertext, and the Remediation of Print, Second Edition.* Mahwah: Lawrence Erlbaum Associates, 2001.

Conners, Robert J. "Landmark Essay: The Rise of Technical Writing in America." In *Three Keys to the Past: The History of Technical Communication.* Teresa C. Kynell and Michael G. Moran. Vol. 7 ATTW Contemporary Studies in Technical Communication. Ablex: Stamford, CT., 1999. 173-195.

Little, Sherry Burges. "Designing Certificate Programs in Technical Writing." In *Foundations for Teaching Technical Communication: Theory, Practice, and Program Design.* Katherine Stapes and Cezar Ornatowski, eds. Vol. 1, ATTW Contemporary Studies in Technical Communication. Ablex: Greenwich, CT., 1997. 273-285.

DESIGN DISCOURSE

COMPOSING

1 The Great Instauration: Restoring Professional and Technical Writing to the Humanities

Anthony Di Renzo

"I hold every man a debtor to his profession; from which as men of course do seek to receive countenance and profit, so ought they of duty to endeavor themselves, by way of amends, to be a help and an ornament thereunto. This is performed in some degree by the honest and liberal practice of a profession . . . ; but much more is performed if a man be able to visit and strengthen the roots and foundation of the science itself."(546)

Sir Francis Bacon, "Preface," *Maxims of the Law* (1596)

Perhaps Giambattista Vico was only half right when he proposed his cyclical theory of history. Besides returning to the same key ideas, civilizations tend to suffer from the same nagging headaches.[1] This is equally true, on a smaller scale, of academic disciplines. They are defined less by their innovations than by their recurring problems and dilemmas.

This paradox certainly applies to professional and technical writing. At the dawn of the new millennium, our discipline faces the same vexing questions it confronted fifty years ago: Are we primarily practitioners and consultants or scholars and teachers? Do we train or educate students? Should we situate our practice in the classroom or the workplace? Is our subject closer to rhetoric and communications or the natural and social sciences?

These questions have become more urgent on college campuses, as professional and technical writing undergoes another turn on Vico's spiral of history. The traditional liberal arts paradigm of higher education is being displaced by a new emphasis on professional and technical training, and emerging PTW programs—especially at small liberal arts colleges—find themselves caught in the middle of the culture wars, simultaneously welcomed and resented, courted and resisted. During this time of risk and opportunity, of breakdown and breakthrough, what is our role and where is our place?

The answer may lie in a Vicoan *ricorso*, a circling back to something old to create something new—a turn-around that is also a turn-about. In the case

---

1. This article originally appeared in *The Journal of Technical Writing and Communication*, Vol. 32. No. 2 (Fall 2002). Reprinted with permission.

of professional and technical writing, this means again proposing that our practice is essential to the humanities. However, I am not simply repeating Carolyn Miller's ideas, already twenty years old, for a more humanistic professional and technical writing practice, much less updating Frank Aydelotte's humanities-centered engineering curricula from the early twentieth century. Instead, taking a cue from Beth Tebeaux's scholarship, I want to suggest returning to the *instructional roots* of our discipline by re-examining the educational ideas of one of its founders, Sir Francis Bacon (1561-1626).

As a scholar and a rhetorician, Francis Bacon straddled three worlds: the literary and philosophical, the administrative and professional, and the scientific and technical—the same mixed audience facing any proponent of professional/technical writing in today's academy. But Bacon is our contemporary in more important ways. Unlike most Renaissance humanists, he located the New Learning (what we now call the humanities) within the related contexts of scientific discovery and invention and professional training and development. Consequently, his proposed educational reforms challenged both the Scholastics, who adhered to the cloistered ideal of the medieval university, and the Ciceronians, who slavishly imitated models of classical rhetoric for imaginary audiences in make-believe situations.

In contrast, Bacon—a believer in public service and the *via activa*—wanted to draw knowledge from and apply knowledge to the natural and social world; and his great treatise, *The Advancement of Learning* (1605), later revised and expanded as *De Augmentis Scientiarum* (1623), is a gigantic curricular blueprint to achieve that end. True education, Bacon argues, should:

- Enhance the professions to make them more ethical, more historically conscious, and more civic-minded.
- Emphasize the material and political conditions of knowledge for the sake of concrete, pragmatic application in the real world.
- Stress the rhetorical underpinnings of organizational and disciplinary discourse, both oral and written.
- Study the media and technologies of science and communications to better government, to reform public and private institutions, and to improve quality of life.

Bacon called his project the Great Instauration, the restoration of true knowledge after centuries of obscurity and neglect, and it went beyond his educational treatises to include his scientific, philosophical and literary works. Updated and revised, Bacon's proposal can be a useful model for creating and defending professional and technical writing programs within the humanities.

To show how, let me gather some of Bacon's educational ideas from his various writings and apply them to the five stages of undergraduate program development: *planning, implementation, mission, design and development, staffing and administration.* Following Bacon's example, I will use aphorisms, since such maxims, he said, force a writer to distill abstract information into concrete principles and to resist the kind of systematic, a priori thinking that shuts down inquiry before one examines the facts.

APHORISMS FOR BUILDING PTW PROGRAMS IN THE HUMANITIES AND SCIENCES

Planning

"He that builds a fair house upon an ill seat, committeth himself to prison."(193)
Sir Francis Bacon, "Of Building" from *The Essays* (1625)

- *To minimize the possibility of failure, construct your program on a solid foundation of research.* Just because you build it, doesn't mean they will come, pace Kevin Costner. Before you draft a blueprint, do some basic marketing. If you already offer one or two basic PTW courses, study their enrollment patterns going back five years minimum and note how these classes fulfill the requirements of outside majors. If you start from scratch, interview departments in the natural and social sciences and the professional schools, determine their academic and professional writing needs and curricular restrictions, and design fitting and responsive courses. These steps will prevent your field of dreams from becoming a bog of screams.

"There are in nature certain fountains of justice, whence all civil laws are derived but as streams; and like as waters do take tinctures and tastes from the soils through which they run, so do civil laws vary according to the regions and governments where they are planted, though they proceed from the same fountain." (287)
Sir Francis Bacon, Book Two, *The Advancement of Learning* (1605)

- *Study the PTW programs of comparable schools, map and analyze patterns of staging and sequencing, then adapt and apply them to your own program.* Use induction to discover the fundamental principles underlying most PTW curricula. Generally, most have *five* stages, each with specific developmental goals and their corresponding courses. For illustration, the following table feature courses from the proposed PTW concentration within Ithaca College's general BA in Writing:

STAGE	GOAL	COURSES
1. Initiation	Use first-year college writing to prepare for professional writing.	WRTG-16300 Writing Seminar: Business WRTG-16400 Writing Seminar: Science
2. Orientation	Teach the building blocks of professional and technical writing at the sophomore level.	WRTG-21100 Writing for the Workplace WRTG-21300 Technical Writing
3. Application	Develop and fine-tune skills through practice and specialization at the lower junior level.	WRTG-31100 Writing for the Professions WRTG-31300 Advanced Technical Writing WRTG-31400 Science Writing WRTG-31700 Proposals, Grants, and Reports
4. Reflection	Frame discipline and practice through history, theory, and rhetoric in upper junior- and senior-level seminars.	WRTG-3600 Composition Theory WRTG-41500 Senior Seminar (PTW)
5. Action	Consult for or intern at an actual company.	WRTG-45000 Internship

Significantly, these stages correspond to Bacon's four divisions of logic and rhetoric in *The Advancement of Learning*: (1) inquiry and invention, (2) judgment, (3) memory, (4) delivery.

"Studies serve for delight, for ornament, and for ability. Their chief use for delight, is in privateness and retiring; for ornament is in discourses; and for ability, is in the judgment and disposition of business." (209)
Sir Francis Bacon, "Of Studies" from The Essays (1625)

- *Be comprehensive.* A hearty education, Bacon believed, should feed the three faculties of the human mind: *reason*, which sees patterns in the world, analyzes data, and posits general principles; *memory*, the mental storehouse of experienced events and material facts; and *imagination*, which channels and articulates the passions and makes intuitive leaps. Even professional and technical training, therefore, should include philosophy, history, and literature.

Implementation

"The ripeness or unripeness of the occasion . . . must ever be well weighed: and generally it is good to commit the beginnings of all great actions to Argus with his hundred eyes, and the ends to Briareus with his hundred hands, first to watch, then to speed." (125)
Sir Francis Bacon, "Of Delays" from The Essays (1625)

- *Although curricular planning should be slow and painstaking, implementation should be relatively swift.* Once you have proposed your program, you are obliged to deliver it. First, create a beachhead to cover your service component, to stake out future development, and to raise expectations. Begin with the nucleus of your projected curriculum, the core courses serving both your majors and outside students, then phase in more specialized classes. Ideally, curricular sequencing should unfold like a paper flower in water.

"As the births of living creatures at first are ill-shapen, so are all innovations, which are the births of time." (132)
Sir Francis Bacon, "Of Innovations" from The Essays (1625)

- *Don't worry, however, if your program assumes a different shape and direction than your original proposal.* Provided these changes are responses to student and institutional need, they indicate evolution not devolution. Being audience-centered and market-oriented, PTW curricula should be flexible and adaptive.

Program Mission

"Expert men can execute, and perhaps judge of particulars, one by one; but the general counsels, and the plots and marshalling of affairs come best from those that are learned." (209)

Sir Francis Bacon, "Of Studies" from *The Essays* (1625)

If your program is housed in the Humanities and Sciences, it should reflect liberal arts values. Unlike PTW programs at polytechnics or research universities, those at small liberal arts colleges should be dedicated less to technical specialization than to what Chase CEO Willard Butcher calls "applied humanities," using the liberal arts to frame and to inform students' future careers (426). A broad base of disciplines and a commitment to civics, Peter Drucker insists, are the best foundation for young "knowledge workers" (5).

"They who have hitherto written upon laws were either philosophers or lawyers. The philosophers advance many things that appear beautiful in discourse but lie out of the road of use, whilst the lawyers, being bound and subject to the decrees of the laws prevailing in their several countries, whether Roman or pontifical, have not their judgment free, but write in fetters. But this task properly belongs to statesmen, who best understand civil society, the good of the people, natural equity, the custom of the nations, and the different forms of states; whence they are able to judge laws by principles and precepts as well as natural justice and politics." (282)

Sir Francis Bacon, Book 8, Ch. 3, *De Augumentis* (1623)

- *Always think socially and institutionally, not only in running your program but in teaching your students.* Professional and technical writing occurs within a nexus of competing discourse communities (business, education, government, and non-profits), and program philosophy, class pedagogy, and curricular design should all reflect that reality. This can be as simple as integrating community service learning into first-year academic writing or as complicated as teaching the classical ideal of the citizen-orator to juniors and seniors.

"Exercises are to be framed to the life; that is to say, to work ability in that kind whereof man in the course of action should have the most use." (118)
Sir Francis Bacon, "A Letter and Discourse to Sir Henry Savile" (1604)

- *Whatever its ideals, your program must provide students with marketable, transferable skills.* Without this "real world" application, your curriculum will be useless.

Curricular Design and Development

"The marshalling and sequel of sciences and practices: Logic and Rhetoric should be used and to be read after Poesy, History, and Philosophy. First exercise to do things well and clean; after promptly and readily." (119)
Sir Francis Bacon, "A Letter and Discourse to Sir Henry Savile" (1604)

- *Provide your students with a clear curricular framework and a coherent disciplinary narrative from the very beginning.* Such context will prevent lower-level courses from becoming too generic and upper-level courses from becoming too specialized.

"Reading maketh a full man; conference a ready man; and writing an exact man" (209)
Sir Francis Bacon, "Of Studies" from *The Essays* (1625)

- *Students should progress from research and analysis, to dialogue and debate, to execution and evaluation.* This curricular staging ultimately benefits all PTW students, whether they choose to become scholars or consultants in the field.

"The mechanical arts, having in them some breath of life, are continually growing and becoming more perfect. As originally invented, they are commonly rude, clumsy, and shapeless; afterwards, they acquire new powers and more commodious arrangements and constructions . . . [till] they arrive at the ultimate perfection of which they are capable. Philosophy and the intellectuals sciences, on the contrary, stand like statues, worshiped and celebrated, but not moved or advanced." (8-9)

Sir Francis Bacon, *The Great Instauration* (1620)

- *Stress tools, not rules.* Since professional and technical writing is practice-driven and context-specific, shun all abstractions. Technology, document design, media dynamics, and institutional constraints should determine your program's curricular philosophy, not the other way around. "Pass from Vulcan to Minerva," Bacon advised (141). Move from praxis to theory. *Never* place theory before praxis. That, Bacon would say, is like building a mansion from the roof down.

"Of the choice (because you mean the study of humanity), I think history the most, and I had almost said of only use." (105)

Sir Francis Bacon, "Advice to Fulke Greville" (1596)

- *Historicize your subject.* That means more than teaching about the development of professional and technical writing. It means tracing the discipline's roots back to classical rhetoric, studying the growth of various social institutions, and reviewing the evolution of different media and technologies. History provides your students with a formative narrative and connects your program to the humanities.

"Histories make men wise, poets witty, the mathematics subtle, natural philosophy deep, moral [ethics] grave, logic and rhetoric able to content." (210)

Sir Francis Bacon, "Of Studies" from *The Essays* (1625)

- *Use case studies to train your students.* Just as young lawyers study past cases to learn legal precedent and to master the conventions and of the courtroom, young PTW practitioners should study past dossiers to learn documentation and to master the demands of the workplace. Case studies are the ideal forum for argumentation and ethical speculation, where students can practice institutional and technological advocacy before multiple audiences.

"There is in human nature generally more of the fool than of the wise; and therefore those faculties by which the foolish part of men's minds is taken are most potent." (94)
Sir Francis Bacon, "Of Boldness" from *The Essays* (1625)

- *Be honest about the politics and absurdity of institutional writing.* Most textbooks skirt this issue by presenting straightforward models and forms and ideal collaborative situations. Your program must address the reversals, rivalries, and irrational thinking that characterize most writing projects and suggest effective countermeasures. At the very least, coping strategies. If you send lambs to the corporate sheering floor, you are guilty of fleecing yourself.

"For it is a rule in the doctrine of delivery, that every science which comports not with anticipations and prejudices must seek the assistance of similes and allusions." (175)
Sir Francis Bacon, Book 6, Ch. 2, *De Augmentis* (1623)

- *Stress the finer points of style and persuasion.* Arrangement, formatting, even striking visuals are not enough to create a winning presentation. Sometimes the telling phrase, the striking metaphor, the provocative analogy carry the day.

"It is a trivial grammar-school text, but yet worthy a wise man's consideration. Question was asked of Demosthenes. *What was the chief part of an orator?* He answered, *Action.* [*Delivery.*] What next? *Action.* What next again? *Action.*" (94)
Sir Francis Bacon, "Of Boldness" from *The Essays* (1625)

- *Aim for results.* "Rhetoric," Bacon claimed, "applies Reason to the Imagination to better move the Will" (238). An effective PTW curriculum will value real-life effectiveness over textbook correctness, which is why you must include credit-bearing internships and consultancies. Seek program feedback, therefore, from potential employers in industry and technology, as well as college administrators, promoters, and admissions officers. And whether or not Bacon actually wrote Shakespeare's plays, make this line from Act 3, Scene 2 of *Coriolanus* your motto: *"In such business action is eloquence."* (79).

Staffing and Administration

"They that have the best eyes are not always the best lapidaries [jewelers]; and according to the proverb the greatest clerks are not always the wisest men." (105)
Sir Francis Bacon, "Advice to Fulke Greville" (1596)

- *Staff courses according to experience and expertise, not seniority and advanced degrees.* This concept seems heretical but makes the best sense and does the most justice to both students and subjects. A full- or part-time instructor who worked for five years as a technical and promotional writer in a county hospital is better qualified to teach medical writing than an assistant or associate professor who graduated from RPI. Scholars can supply practitioners with outside readings, but practitioners cannot supply scholars with inside knowledge.

"Surely ever medicine is an innovation, and he that will not apply new remedies must expect new evils. For time is the greatest innovator, and if time of course alter things to the worse, and wisdom and counsel not alter them to the better, what shall be the end?" (132)
Sir Francis Bacon, "Of Innovations" from *The Essays* (1625)

- *Anticipate change and plan for contingencies.* To keep your program open and flexible, be prepared to alter its focus and sequencing and to amend, combine, or jettison courses in response to market need and student demand. On the subject of adaptability, Bacon loved to quote Machiavelli: "If you can change your nature with times and circumstances, your fortune will not change" (68).

"The proceeding upon somewhat conceived in writing doth for the most part facilitate dispatch; for though it should be wholly rejected, yet that negative is more pregnant of

direction than an indefinite, as ashes are more generative than dust." (135)
Sir Francis Bacon, "Of Dispatch" from *The Essays* (1625)

- *Compost your failures to fertilize future projects.* Recycle rejected courses as special seminars. Transplant background material from an aborted proposal into a program report. Boilerplate unread course descriptions when submitting a catalog copy. Waste nothing.

"Just as some putrid substances like musk or civet yield the best scent, so base and sordid details sometimes provide excellent light and information." (122)
Sir Francis Bacon, Book One, Aphorism 120, *The New Organon* (1620)

- *Even when things stink, welcome confusion and disappointment.* If you can bear the temporary din of frustration, your program's elements eventually will harmonize. In science as in music, Bacon said, dissonance is necessary to fine-tune an instrument.

A Baconian approach to curricular design and implementation offers three distinct advantages to emerging PTW programs at small liberal arts colleges. First, Bacon's educational principles and practices make a convincing apologia for most English departments and writing programs. The Lord Chancellor is the best lawyer to plead your case because he appeals to so many different audiences. Traditional humanists will be pleased to see how Bacon's ideas about professional and technical writing fit historically within their own disciplines. Theorists and New Historians will respect his materialism and praxis, while department chairs and program directors will appreciate his shrewdness and practicality.

Second, Bacon's pragmatism and social conscience wed humanistic education to public policy and public works. As both a legislator and a jurist, James Spedding observes, Bacon "could imagine like a poet and execute like a clerk of works," qualities that will appeal beyond a department's curriculum committee and will engage college administrators and representatives from research and industry (72). Bacon was committed to achieving concrete results in the real world. His *summum bonum* was the social good. Indeed, as J. G. Crowther explains, Bacon believed "the most determined statesmen are those who are deeply versed in social philosophy, and are engaged in carrying out policies based on a profound study of the principles of nature and society" (44). Small, liberal arts colleges should adapt this philosophy in their humanities-based PTW programs, using professional and technical training to bridge the gap between the quad and the commons.

Last, Bacon's radical rethinking of the sciences and the professions can inspire programs to re-imagine their pedagogy while providing the necessary theoretical scaffolding to paint the big picture. The loam of historical research can provide rich soil to grow good programs. Bacon, an avid gardener and landscaper, makes this analogy in Book 6, Chapter 2 of *De Augmentis Scientiarum*:

> For it is in arts as in trees—if a tree were to be used, no matter for the root, but if it were to be transplanted, it is a surer way to take the root than the slips. So the transplantation now practiced of the sciences makes a great show, as it were, of branches, that without the roots may indeed be fit for the builder, but not for the planter. He who would promote the growth of the sciences should be less solicitous about the trunk or body of them and lend his care to preserve the roots, and draw them out with some little earth about them. (172)

However, we scholars and teachers of PTW should look back less to legitimize our practice for the sake of our critics than to look around and look ahead for the sake of our students. Bacon was no antiquarian, after all. Although he venerated history, he believed people should use the past primarily to secure present provisions for a future journey. The frontispiece of the 1620 edition of *The Great Instauration* shows a billowing galleon returning through the Pillars of Hercules from its voyage on unknown seas. If the latest turn in the academy has made our discipline more valuable and necessary, if it is now our turn to define the rules of the game, if this collective return to our intellectual past is to be more than academic, then we must recapture our sense of wonder with our sense of mission. In T. S. Eliot's words:

> We shall not cease from exploration
> And the end of all our exploring
> Will be to arrive where we started
> And know the place for the first time. (59)

WORKS CITED

Bacon, Francis. "The Advancement of Learning." *The Oxford Authors: Francis Bacon.* Ed. Brian Vickers. Oxford: Oxford University Press, 1996. 120-299.

Bacon, Francis. *Advancement of Learning and Novum Organum*. Revised Ed. Ed. and Trans. James Edward Creighton. New York: Colonial Press, 1899.

Bacon, Francis. "Advice to Fulke Greville." *The Oxford Authors: Francis Bacon*. Ed. Brian Vickers. Oxford: Oxford University Press, 1996. 102-106.

Bacon, Francis. *The Essays*. Ed. John Pitcher. New York: Penguin, 1985.

Bacon, Francis. *The Great Instauration and New Atlantis*. Ed. J. Weinberger. Arlington Heights, IL: Harland Davidson, 1980.

Bacon, Francis. "A Letter and Discourse to Sir Henry Savile." *The Oxford Authors: Francis Bacon*. Ed. Brian Vickers. Oxford: Oxford University Press, 1996. 114-119.

Bacon, Francis. "Maxims of the Law." *The Works of Lord Bacon*. Vol. 1. Ed. James Spedding. London: Henry G. Bohn, 1854. 546-590.

Bacon, Francis. *Novum Organum*. Ed. and Trans. Peter Urbach and John Gibson. Chicago: Open Court Publishing, 1994.

Butcher, W. F. "Applied Humanities: They Will Pay You for the Other Five Percent." *Vital Speeches of the Day* (1 August, 1990): 623-25.

Crowther, J. G. *Francis Bacon: The First Statesman of Science*. London: Cresset Press, 1960.

Drucker, P. F. *The Effective Executive*. Harper, 1966.

Eliot, T. S. *The Four Quartets*. New York: Harcourt, 1971.

Machiavelli, N. *The Prince*. 2nd Ed. Ed. and Trans. Robert M. Adams. New York: Norton, 1992. Shakespeare, William. *Coriolanus*. Ed. Jonathan Crewe New York: Penguin, 1999.

Spedding, J. *Francis Bacon's Personal Life Story*. London: Rider and Company, 1949.

2 Starts, False Starts, and Getting Started: (Mis)understanding the Naming of a Professional Writing Minor

Michael Knievel
Kelly Belanger
Colin Keeney
Julianne Couch
Christine Stebbins

INTRODUCTION: NAMING AS RHETORICAL DISCIPLINARY/PROGRAMMATIC ACTION

After several years of planning and development, the University of Wyoming Department of English now offers an undergraduate minor in professional writing. In thinking about our program, we have become increasingly conscious of the ways in which the name of this program, simply the "professional writing minor," functions within our institutional context, a relatively small (approximately ten thousandundergraduates) state university and a traditional English department offering both undergraduate and graduate (MA and MFA) degrees.

All programs have names, but most, including our own, are not particularly noteworthy. Save for some notable exceptions in recent years (for instance, Central Florida's doctoral program in "Texts and Technology"), most writing programs that identify their mission as distinct from composition or creative writing, regardless of size or status, rely heavily on a familiar word bank for their program titles: "rhetoric," "communication," "writing," "technical," and "professional." But while this uniformity has helped fashion a quasi-recognizable disciplinary identity in "nonacademic" writing and communication, it also deflects attention from the significance of signification. Awash in the hundreds of questions and issues that come with envisioning a program, teachers and administrators may move uncritically past this vital step in the development process, reaching for terms in the word bank without sufficiently considering their implications and the multiple lenses through which those words will be read.

Much, it seems, is at stake when naming a program. Robert Johnson points to a name's ability to make things "unforgettable"; however, he acknowledges that the process of naming is complex and fraught with competing motives, asking, "Is the naming of programs a determinist enterprise that takes on a life of its own? Or are we being creative in our endeavor to associate thing to thing, spiritual fact with embodied form?" Johnson recognizes the need to let local factors guide naming but cautions against promising more (or less) than can be delivered: "…should we think twice about *unnaming* ourselves in the process of trying to embrace too much?" Generally speaking, the implications of program naming have been inferred from broader conversations about connections between program development and institutional politics (Cunningham and Harris; Hayhoe, et al; Latterell; MacNealy and Heaton; Mendelson; Rentz; Sides; Sullivan and Porter) and intersections between disciplinarity and professionalism (Faber, Savage).

With their focus on larger programmatic and disciplinary issues, many of the aforementioned authors typically address program naming in tangential fashion, although some acknowledge what might be at stake when naming a program or, in some cases, an entire field of inquiry. MacNealy and Heaton suggest that the name "Professional and Technical Communication" may best represent the field's scope and hope for acceptance: "…if we want to enhance our image among those outside the field, the term 'professional' might be a better choice than 'technical' because it is more inclusive and it sounds less mechanistic." (55). Dayton and Bernhardt's 2003 survey of ATTW (Association of Teachers of Technical Writing) members asked respondents what the field should be called, offering a variety of fixed-response possibilities from which to choose. The top three choices included: "Technical Communication" (39%); "Professional Communication" (32%); and "Professional Writing" (10%). However, in an open-ended follow-up question, respondents offered still more alternatives and noted the importance of having a name that communicated clearly to outsiders but that acknowledges specific contexts (29-30).

We know, then, that naming—of the discipline, of programs—is a contested process. But beyond being a critical choice in the early stages of a writing program, we believe that a program name is a powerful site from which to begin examining a program's history, politics, and function—a program name tells a compelling story. We argue that any study of naming becomes, in part, a study of 1) historically-situated program development, and 2) program execution, one test of a name's veracity and scope, as well as the implications of its signification. Thus, in this chapter, we trace the development of the professional writing minor at the University of Wyoming through a narrative chronology that constructs a constellation of the voices (writing faculty, other English department members, administrators, and students) giving shape to the minor as it currently stands; specifically, we examine our "starts" and "false starts" before turning to the present challenges of "getting started." In doing so, we map the vast array of connected and disconnected questions, concerns, and values that come into play when a program of this kind is developed and named. We believe that the archaeology of a program name can be uniquely generative as a site of research, a catalyst for institutional critique, and, consequently, a means of reclaiming a name and program. And while we acknowledge the power of more abstract conversation about naming, we assert that a local focus might yield more granular insight into this highly contextualized process, insight that has the potential to enrich—and complicate—our sense of the complexity of both naming and program development.

FINDING OUR OWN VOICES: WINDOWS TO PAST, PRESENT

In approaching the question of program naming, we prioritized the two broad currents identified above: 1) historically situated development and 2) program execution. To that end, we crafted a quasi-ethnographic approach to researching our name and the issues and events that both precipitated and emerged from it. In short, we compiled information and perspectives through examination of:

- our own personal narratives written from the perspective of writing faculty deeply invested in planning, teaching in, and overseeing the program
- semi-structured interviews with past and present members of the English Department (faculty, students, administrators), many of whom played an integral role in the development and launch of the program
- files and archives containing a variety of documents pertaining to the minor (e.g., course approval forms, meeting minutes, related grant proposals, email correspondence regarding the curriculum, computer classroom, etc.).

As writer-researchers, we represent both a historical cross-section of the writing history at UW and the range of responsibilities for program execution at our university. All of us are situated in the Department of English. Some of us work as academic professional lecturers (APLs), which are extended-term teaching positions (six-year renewable appointment and opportunity for promotion). Others are assistant and associate professors, respectively, in writing-related fields.[1] Some of us have a significant measure of professional writing experience outside the academy in addition to experience in other fields; others have focused more specifically on writing in academic contexts. All of us have taught a variety of courses in our department's professional writing minor, served on a range of writing-related committees, and worked together on various writing-related initiatives in our department or on campus.

At UW, we have constructed a minor designed to capitalize on the range of experience and expertise that we, as teachers, bring to the program. At present, the professional writing minor consists of eighteen credit hours and emphasizes flexibility. Students are required to take two three-credit core courses:

ENGL 2035	Writing for Public Forums
ENGL 4000	21st Century Issues in Professional Writing

In addition, they choose two of the following three-credit courses:

ENGL 4010	Technical Writing in the Professions
ENGL 4020	Editing for Publication
ENGL 4050	Writer's Workshop: Magazine Writing
ENGL 4970	Professional Writing Internship

Finally, students select two writing-intensive elective courses, typically related to their major course of study and connected to their career objectives.

CHRONOLOGY: CONSTRUCTING OUR PAST, CONSIDERING OUR PRESENT

In the sections that follow, a series of narratives describes the myriad conditions, values, and beliefs that gave rise to a program named, somewhat serendipitously, the "professional writing minor" and demonstrates some of the consequences of this naming choice for various stakeholders within our institutional context.

Starts (1986-1993)

It would be inaccurate – and unfair – to suggest that nothing occurred toward writing development at the University of Wyoming prior to 1986. Tilly and John Warnock began their careers at UW during the 1970s and their impact lingers to this day. Of writing at UW and across the state, one colleague recalls, "I think it was an outgrowth of the Warnocks ... they were a major, charismatic force in the department, (and) not just within the department but in the university as a whole." Another colleague recalls their development of the Wyoming Writing Project, the Wyoming Conference, and the Writing Center during the seventies and early eighties. Their collaborative essay, "Liberatory Writing Centers" (1984), both defined and helped establish university writing centers nationwide, and Tilly's *Writing Is Critical Action* (1989) is still commonly cited in composition scholarship. In essence, the Warnocks were the first real representatives of composition and rhetoric—as we would define that discipline today—at UW, and were strident advocates for its acceptance.

The late 1970s also begat a pivotal course on campus: Scientific and Technical Writing (ENGL 4010), the name of which, interestingly, would be changed to "Technical Writing in the Professions" in 2001. As shall be seen, tracking 4010's permutations constitutes a primary, connective thread through our narrative. If nothing else, one colleague notes, "I'm sure that (4010) proved the existence of a clientele" for an upper-level writing course beyond that era's requirement for only two semesters of "freshman" composition. Twenty years later, meeting the needs of that "clientele" would, in part, spawn the professional writing minor.

On the other hand, the advent of Scientific and Technical Writing almost immediately raised two counter-considerations. The course was developed within the English department from a direct request by the College of Engineering – to enhance their students' writing skills – but the College of Business quickly came onboard and began requiring it of their majors. For obvious reasons, the course was immediately consigned to the "service" bin, with the result that very few English faculty members cared to teach it. This attitude was administratively underlined when the Dean of Arts and Sciences subsequently refused to accept work in this area for tenure or promotion deliberations. Because of this, and because the course was too advanced for graduate assistants to teach, 4010 was progressively shunted to temporary lecturers.

And then there was that *name*—"Scientific and Technical Writing." Clearly, when marketing or accounting majors began queuing up for the course, it lost any technical edge or scientific facet it might have contained. Indeed, one faculty member who developed the original version of 4010 thought to himself, at that time, "This really isn't a scientific and technical writing course ... we ought to call it 'professional writing.'"

This brings us to our primary timeline from 1986 to present; we chose 1986 as a starting point for one simple reason: that year, two hundred attendees of the Wyoming Conference on English (co-chaired by the Warnocks) overwhelmingly adopted the "Wyoming Conference Resolution," arguably the most important document concerning post-secondary writing in our professional lifetimes. With its focus on personnel issues, today the Resolution seems akin to a union's grievance against management. However, by concentrating on *people*—on those who teach and develop writing—the Resolution served as a cornerstone for comprehensive writing curricula across the country. Indeed, the Resolution helped make it possible to develop writing curricula by emphasizing improved working conditions, such as compensation and workload, for those who would develop and execute such programs. But as we now know, few of these achievements came smoothly or without some sort of price, and writing development at UW was certainly no exception.

Without fanfare – and with virtually no attention from other department members – our assistant chair began a "cohort group" for 4010 instructors in 1987. The group's initial function was twofold: to supply mutual support for those teaching this demanding course, and to improve consistency without limiting academic freedom. The cohort group's overall success was confirmed by one colleague who joined the department a few years later: "The group… seemed to feel a justifiable sense of ownership of the course and pride in its high quality and had reached a (general) group consensus on standards and assignments." Certainly these were no small accomplishments, but they frequently played second fiddle to larger topics within the group. For instance, for several years, the group maintained a running discussion of gender issues in the technical writing classroom, such as why male instructors were often evaluated as being "tough but fair," whereas our female colleagues were raked for being "too tough," "unfair," or "a bitch." (Combined with being stuck in term-limited positions, teaching a devalued course, and working in an "unscholarly" discipline, this gender bias formed what one colleague dubbed a "quadruple whammy.") Under the circumstances of the times, it was invisible work performed by an invisible group, but it "… solidified and brought together the APLs (lecturers) in the department who were working with 4010."

More visible by far were the events of 1990-93 and the English department's response to them. First, UW's administration mandated development of a new University Studies Program (USP), and central to that plan was replacing the previously mentioned two-semester "frosh comp" requirement with writing courses labeled WA (first-year), WB (sophomore/junior), and WC (senior/capstone). After review and approval, any college, department, or program on campus could teach any of these multi-tiered writing courses. The English department reacted by appointing a six-person Writing Committee and charged this group with qualifying, quantifying, and separating these different levels of written discourse.

All of this was rather momentous. The new USP simultaneously recognized writing's central importance to a meaningful education *and* opened the door to writing across the curriculum (WAC) for the first time at UW. In the eyes of the university's administration, at least, those who taught writing were suddenly elevated from second-class citizenship to being significant contributors. And while the Writing Committee's official function was to determine what constituted WA, WB or WC writing only within this department, it was tacitly understood that our delineations ultimately would apply to all writing courses, campus-wide. One lecturer remembers, "We were considered 'the pros' when it came to writing, so we got to call the shots." Therefore, through the act of defining, this small in-house group *named* writing at UW.

This section would be incomplete without mentioning a Department of English retreat held in the fall of 1993. This gathering produced the departmental decision to formulate a "writing program," that focused on neither "academic" nor "creative" writing at its core and sparked the need for someone to develop and direct such a program. However, individual recollections of this event are varied. One participant remains convinced that this portion of the retreat's agenda was orchestrated to the point of crafty manipulation ("… it was a nifty bit of stacking the deck"); two others would contend all of this "just happened" with little to no forethought or planning; and at least one department member can recall precisely who catered the food – and nothing else. One might suspect that the clarity and tone of these memories depended on the individual's proximity to writing and writing instruction, but that could be mere conjecture.

False Starts (1993-1998)

By the end of this period of "starts," the value of Scientific and Technical Writing (ENGL 4010) was clear on paper, at least regarding numbers, as evidenced by a University Studies document authored in part by the English department chair in 1991. This document focused on the fact that freshman composition and ENGL 4010 made up most of the department's course offerings and helped keep the department viable in the eyes of the rest of the University; indeed, in the eyes of our department chair at the time, 4010 helped "justify its [the English department's] existence and size to the outsider." Thus, the worth of these two writing courses in the larger university context was becoming clearer.

The aforementioned WAC movement of the early 1990s played to mixed reviews campus-wide but had significant implications for the APLs charged with much of its implementation, as well as for the way in which the department was perceived vis-à-vis writing on campus. A former chair, now a dean, believes that the department's involvement in WAC showed that "…we in English are 'good soldiers'" to the university at large. She also believes that because of WAC our writing teachers got more respect campus wide because of a heightened presence, if not necessarily in our own department. For Writing Center personnel, nearly all of whom were English department APLs, these WAC-focused years were busy. In addition to full course loads, most APLs were assigned to the Writing Center for five hours a week to work with clients and perform extensive outreach for the Center, often preparing and presenting numerous workshops and seminars each week to help guide the campus-wide implementation of WAC. In the end, however, APLs could claim little if any meaningful professional credit for this tremendous outlay of individual and collaborative time and effort; it was just expected. Ironically, but politically foreseeable, it was the relatively invisible, relatively powerless temporary writing instructors who were charged with helping to improve the level of writing integration in the entire university.

When some WAC courses around campus were later dropped, departments typically directed students to 4010 to meet graduation requirements, and so course enrollments continued to burgeon. However, some English department faculty felt that this type of writing was too far outside the domain of traditional English Studies and a threat to the very identity of our department. In consequence, English majors were not allowed to take the technical writing course for credit in the major. One senior lecturer says, "The problem we've always had with the perception of 4010 is that people always saw it as a service class for people outside of the English department and of course, as you know, it wasn't allowed to be counted for an English major… people saw it as being like fill-in-the-blank kind of writing and I guess they didn't see it as "real writing" … they just saw it as a real sort of pedestrian writing." Another faculty member notes, "…the course…has always had this marginal relationship to the department. I mean, it was so striking and odd to me that for a while that course didn't count toward the major …that was one 4000-level course that 'non-professors' …could teach."[2]

While ambivalence toward the role of technical and scientific communication remained, the department moved to build upon its decision at the 1993 department retreat to start a real writing program at UW by making a professorial hire at the assistant level in rhetoric/composition in 1994. After a honeymoon year to "get her feet on the ground" the department expected the hire to open a new chapter in our freshman writing program, especially in the development of teaching assistant (TA) training, as graduate TAs taught many of the composition courses. Unfortunately, the professor's perceived overall resistance to guiding the program and her self-confessed anger at the political situation regarding the overall attitude toward writing resulted in her resignation after two years. In her time here, however, this professor helped lead the technical writing instructors more fully into the world of computer technology and computer-mediated instruction through workshops and training sessions with her and outside consultants.

In the ensuing two-year gap between the departure of one rhet/comp professor and the hiring of another, there seemed to be growing consensus regarding the need for a "tenure track presence... to give a new writing program legitimacy." Throughout this tough time, the technical writing cohort hung together, trying to keep spirits up, lives intact, and eyes looking forward as the professionals that members knew they were. The cohort kept abreast of new trends, technology developments, and the national debates about the many aspects of the discipline. The one thing members did not formally discuss, however, was a professional writing minor. Although the 4010 cohort group would later play a central role in constructing the minor, at this juncture, it was just "too pie in the sky" to have any real hope it might happen.

Getting Started (1998-2000)

However, in the October 1997 *MLA Job Information List* the UW English Department publicly indicated its intention to develop a writing minor and sought a senior faculty member to serve as a "point person" for the new minor and the first-year writing program. The department's intention to hire at the senior level indicated an awareness—born during the years of "starts" and "false starts"— of the political complications inherent in coordinating or developing writing programs within a department holding a traditional literature view of the English Department's curricular geography (Sullivan and Porter 393). One senior literature professor, to whom a former department chair attributes the idea of developing a writing minor, also points to a generational shift in the department in which a cohort of faculty "came out [of graduate school] with a much different notion of what "English" meant for our students, and not just students who were going to show up in our English classes because of their great love of literature, but students who were actually living and working in English." She explained in an interview that "for us, thinking about writing as a part of a student's education wasn't an add-on. We saw the integration." She believes this integrative vision among some faculty members paved the way for the 1998 hiring of an Associate Professor of Composition and Rhetoric and for a significant store of goodwill among the literature faculty toward a possible new writing minor.

Perhaps understandably, considering the departmental history and hierarchies, in the early stages of developing the minor, some of the department's lecturers were more wary than their literature colleagues about the hiring of someone who, although appropriately credentialed with a doctoral degree, had less nonacademic experience with technical and professional writing than most of them did. The new hire, Kelly Belanger, brought to the position a generalist background and interest in collaborative program development that proved a relatively comfortable fit with a department whose literature and creative writing faculty had only a nascent sense of composition, rhetoric, or professional/technical writing as fields within English Studies, each with their own bodies of scholarship and intellectual traditions. Along with research interests in composition, computers and writing, business communication, and literature, she also brought entrepreneurial experience from having developed a new writing program for unionized steelworkers in Ohio and, with business partners, a coffeehouse/café. This generalist background to some extent mirrored the generalist strengths of the department's richly experienced APLs. Even so, early on, some members of the 4010 cohort greeted the new "point person" with skepticism that made it difficult for the team and their appointed leader to see their common interests in advancing the status of writing in the department. One senior APL proved a valuable intermediary, "translating" between other APLs and thus helping to clarify their overlapping goals. As the longtime leader of the cohort group put it, "I think we had the perception that something like [a writing minor] couldn't happen."

Although we can't identify the particular meeting or discussion during which we settled on the term "professional" to characterize the writing minor—indeed it seemed a name simply "in the air" that we gravitated toward—notes from a June 2000 Wyoming Conference on English writing workshops suggest that some members of the technical writing cohort group pondered early on the implications of the term "professional." One note taker mused, "Professional writing is an umbrella term. Business writing/com, tech. writing, and scientific writing are all subsumed under the larger term 'Professional Writing.' Which of these terms work best for what we want to teach?" A senior APL explains, "We weren't trying really to narrow our program because first of all we're all kind of generalists.... And I think that we felt comfortable with a more general name, or general title, under which we could see ourselves as instructors. Professional writing minor seemed just right."

The scope of the minor broadened even further when the proposal for the minor went before the department in Spring 2000. Literature colleagues argued for including literature, creative writing, and any designated WAC courses as electives in the new program. These arguments reflect what a former department chair identifies as the initial impetus for the minor when it was first discussed during the 1993 retreat—to draw in more students to the English Department, including its literature courses. Rather than debating the boundaries of the minor or exploring what benefits clearer articulations of what the minor courses could offer intellectually as well as practically, the department agreed upon a big umbrella for the minor and moved forward quickly to approve it.

Settling on the term "professional," with its ever-expanding connotations, not only reflected the generalist background of faculty teaching in the minor, it also responded to a range of desires, anxieties, and assumptions on the part of the English Department and its faculty. While some, even many, faculty members might have welcomed more sustained discussion at least behind the scenes, the department appeared willing, even eager, to approve the minor without further discussion, perhaps for practical reasons of its own. Perhaps anxious not to go the way of impoverished, diminished humanities departments with no service course responsibilities, some faculty saw the new "professional" writing minor as a commodity to package and sell, a product more practical and marketable than its literature or creative writing courses. One colleague described using the term professional as a "packaging maneuver." And in the early 1970s, teaching technical writing courses had seemed a wise career move for one literature professor we interviewed, who feared for his career in light of declining English majors. Another literature professor interviewed denied that his support for the minor had anything to do with concern about the viability of the English Department or major. Instead, he saw the minor as a way to address the perceived illiteracy of engineers and agronomists while potentially drawing them to take a few literature electives and the "richer experience" those courses offered. But his quick denial of any concern about English Department enrollments belies the reality that more majors and minors translates to more faculty hires, a larger budget, and more influence for the department in the university.

Only one literature professor strongly expressed concern about "the validity of a Writing Minor in the first place." When the proposal for the minor went before the department's curriculum committee, few wanted to debate questions he raised about whether "the minor value[s] praxis above the quest for pure knowledge" and whether "'writing' as defined by the minor represent[s] a field of knowledge or a set of skills?" The discussion closed down quickly after a counterstatement claiming that "as far as praxis goes, schools like Engineering are already structured around the concept of praxis. As well, elements of our literature courses can be considered to be skill-based." Minutes from the meeting record that "we (committee members) did not resolve disagreements on this issue." More to the point, the brief discussion begged the question at the heart of the matter—whether courses in writing, rhetoric, and communication are legitimate areas of intellectual study in a research university or whether these courses and those who teach them merit the adjunct and secondary status constructed for them by the English Department's curricular geography.

Unfortunately, much of the intellectual work that took place in creating the professional writing minor remained invisible to the literature faculty and even to the department chair at the time. The former chair recalls that "somebody might have mentioned to me that Kelly and some of the others – the academic professionals, I guess, were talking about [a writing minor], but I don't recall ever getting anything official you know, and I thought, let the discussion go. People should talk about things like that." What the faculty didn't see, or caught only glimpses of, was two years of intensive work that involved three major grant projects: one to develop a computer classroom, another to develop a sophomore-level WAC course into a foundational course for the new minor, and a third to develop the minor itself through a Center for Teaching Excellence (CTL) Grant that funded a retreat, a series of workshops, and an assessment survey of the English 4010 course. A senior APL remembers the English 4010 survey project as "the first time we were really coming together as a group of 4010 teachers and realizing that we had something that was a fairly good course—very important to the university—and the fact that we were teaching most sections meant that it was very important to the English Department, even though our majors weren't eligible for it."

Meanwhile, the CTL grant application reflects the cohort group's determination that courses in the minor be intellectually rigorous, grounded in relevant discourse theories:

> As part of our Academic Plan, the Dept. of English is developing a new, interdisciplinary minor in Professional writing. The minor will prepare students from a range of disciplines for writing-related careers and deepen their understanding of the social, political, linguistic, and rhetorical nature of written discourse.

Despite this "mission statement" avowing the intellectual underpinnings of courses in the minor, we've learned through this project that some of our literature colleagues remain unconvinced of the minor's place within a liberal arts tradition. As one colleague observed, "a lot of us on the lit faculty had a general sense of things that went on in professional writing overall but no sense of the kind of intellectual history or the intellectual debates animating the field. I think a lot of people didn't even know it existed because they thought it was more of a 'toolkit' minor." She added, "I think that was a real failure on our part for a long time in the department to even recognize that there is an intellectual history to this stuff, not just…it's not like becoming a mechanic or wrenching or something…it was an odd…it was a slow education for me." In truth, the failure has been mutual. In our zeal to "get started," those of us working on the minor underestimated the challenge and importance of bridging the gaps between our own and our colleagues' understandings of what professional writing and, more broadly, rhetorical education and scholarship can entail. As long as this gap remains, the future of the minor rests on unsteady ground.

Staying Afloat (2000-present)

Navigating the waves of resistance and tides of support, we have reached a relatively calm harbor where our minor rides nicely in the water. We have a strong set of courses that are well received by students, taught with competence and creativity by our tenure-track and APL faculty. We continue to teach most of our courses in the Humanities Resource Center, more commonly called a computer classroom, housed in the building in which the English department resides.

The minor attracts students from across campus who tailor their elective choices to match their career expectations or their interests. At present we are unable to articulate, exactly, what it is the professional writing minors as a group expect as a career payoff for their efforts. However, many of us have heard students express comfort with the title "professional," as it connotes what they've studied, not what it leads them to expect. Students don't seem burdened with questions of semantics and what role word choice plays in our department's administrative plans, staffing decisions, or interaction with the university as a whole. For this reason, our minor could largely be labeled a success: we have departmental support, strong collegiality among writing faculty, and student enthusiasm. What more could we want?

We would not be doing a service to our various constituencies if our answer to that question was "nothing." Instead, we continue to seek clarity in our goals and objectives in order to foster departmental and university support. We know the educational bazaar represented by students who chose our minor is going to mean we'll have to deal with various expectations on the part of students and their major departments. One teacher reports that in her recent 2035 class (the introductory course in the minor) she had computer science majors, physics majors, a business major, and somebody in biology or botany or forestry, and "all of them were interested in the professional writing minor. Their major doesn't have that focus, but whatever they end up doing, they like to write and really want these skills." One English major decided to declare the minor because "I had no idea what I wanted to do with my degree. It (the minor) seemed to diversify my choices, instead of being limited to just teaching, which is what you often hear is 'the only thing you can do with an English degree.'"

Faculty expectations of the minor vis-à-vis student outcomes also vary. One colleague suggests that the minor might "get them [students] jobs, and get them prepared for those jobs. It would get them official certification that they could carry out into the world that they had not only done some writing, but that they'd thought about it and gotten some training in thinking about writing for a variety of contexts in the professional world." But others in the department are still uncomfortable that English is in the business of getting students ready for jobs. States a former department chair:

> The worry that I had was that by creating this analogy (calling the minor "professional") that there would be an implicit promise of where you go through this and there is a profession of technical writing and you can – you will – get a job in it. And I began to worry that students would in a sense get the idea that they were entitled to a job.

He goes on to support some form of employability in the majors and minors we turn out. He sees the role of the minor as:

> ...producing writers capable of learning to write software manuals, or capable of learning to write grant proposals for whatever agency, or capable of learning to write contracts as paralegals. I don't mean that you'd have a minor in writing software manuals. What you would have is a certain fundamental grounding and awareness of writing software manuals, that there are certain conventions in legal writing, there are certain conventions in grant writing....

Uncertainty over how we should prepare students inevitably spills over into how we should hire teachers to do the preparing. In 2002 we hired a second tenure-track rhet/comp person expected to be a major player in the minor, and in 2004 we hired an APL expected to teach 4010 through our Outreach school and run the professional writing internship program.

One tenured literature professor who was involved in searches for a second tenure-track rhet/comp position phrased the uncertainty like this:

> ...you could tell with each potential candidate, the meaning of what the professional writing minor would be would change because it would be, 'oh, here's somebody who's amazing at web design, and that's what our professional writing minor needs.' But then it would be, 'Here's somebody who does science studies. That's what the professional writing minor is.'

During that process, which took place before we began this research, these sometimes contradictory senses of need were not sufficiently recognized or articulated. The analogy of people grasping different parts of an elephant and not understanding they all have the same creature in their hand is oft-used for a reason.

Our minor accommodates many interests, but that part which we each claim as our own causes us to claim stakes in a territory that is only part of the whole.

The first year of our search for the second rhet/comp position produced no job offers, but in the second we found a candidate who fit what we could agree were our needs. Michael's degree was in technical communication and rhetoric, and he was looking for something that would allow him to pursue that interest. He explains what he expected from a program called "professional writing" and how that compares with what he found: "I saw a 'professional writing' minor and immediately associated it with technical writing—the terms are/were often used nearly interchangeably in the professional literature; it was my hope that I could participate in the minor by finding a technical communication niche." The uncertainty the department felt was in place not only for the hiring of tenure-track professors. Indeed, in the course of the recent APL hire, we continued to disagree about what our needs were and which job candidates best met them. Should we value technical skills? Broad training in rhetoric? Professional writing experience? In each case, the lack of consensus regarding the name "professional writing" turned the job search into a heuristic for understanding the field and our program.

Some final thoughts as we reflect on where we are at this time. We've realized through this research project that what started out as an English department service course (4010) has turned into a viable and exciting minor for our students. In turn, our APLs and tenure-track faculty have a stronger sense of professional purpose. While it is true that the naming of the minor raised issues for debate, that debate is a healthy one because it takes place among an increasingly respected writing faculty within the context of the larger department's view of its purpose and identity. The result is better integration of our once-perceived disparate needs into a department that, for now, at the worst, accepts what we do and at the best, celebrates it.

UNCOVERING A HISTORY OF NAMING: LOCATING KEY THREADS

We started this project with a key assumption: naming itself is a generative process, and examining a name and its historical context can yield important insights about that context and the way in which a name functions. While we remain somewhat unsure of the exact moment in time when "professional writing minor" was minted as the program title, by examining our history, we are able to reconstruct not only a rationale for the name but also the significance of the naming context—the politics, decisions, attitudes, and actions that spawned the development of the program itself. We recognize, too, the process of naming and investing a name with meaning as a claiming of sorts. Constructing a history of our name permits a reclaiming, an opportunity to look back and move forward mindful of this heritage.

History and Development: A Service Course Heritage

What does our history tell us? For one, our "professional writing minor" exists in a context where writing has, in some ways, been historically marginalized or, worse, de-professionalized. We can see that in our particular institutional context, professional writing has a lineage traceable to the early days of a politically charged course, ENGL 4010. As a service course at the heart of the department's commitment to writing on campus, it has long generated credit hours and revenue for the department but has not always been fully embraced. Given its role as the department's most prominent non-creative, non-literary writing course outside of freshman composition, 4010's history no doubt conditioned the development and perceptions of the professional writing minor. As noted earlier, the idea of a minor or writing program of some sort fermented informally for years in the 4010 cohort group, and it seems possible that the minor's historical connection to 4010, while lending credibility to the minor campus-wide, may now consciously or unconsciously compromise how members of our department see it and its function within the department. It is difficult to ignore the fact that advanced non-creative writing study was traditionally offered only through 4010, a *service* course with all the attendant baggage of such purpose. By association, the minor—which includes 4010 as an elective—cannot help but evoke thoughts of service and utilitarianism. Consequently, the specter of "toolkit" and its associations with intellectual—even moral—bankruptcy looms.

Service courses, as we know, are frequently taught by adjuncts and lecturers whose job security is oftentimes in question, and the residue of 4010's service history lingers. At the center of the 4010 story, and by association the professional writing minor, is the group of people who have traditionally taught it: extended-term and temporary academic professional lecturers (APLs). While APLs gained presence and influence around campus in the mid-1990s with the emergence of WAC on campus, much of their work was done through the Writing Center (housed outside of English); thus, much of the APLs' work was rendered largely invisible, which, in turn, failed to raise their profile or the profile of devoted writing instruction in the English department. The staffing situation was only further complicated by the elimination of funding previously earmarked (in the wake of the Wyoming Resolution) for additional extended-term APLs; when these funds disappeared, the department turned to a growing roster of temporary lecturers working on one-year contracts. Until Kelly arrived in 1998, then, the segment of the faculty with the least "professional" status in the traditional academic hierarchy—regardless of teaching success or professional writing experience—conducted the vast majority of advanced writing instruction in the department.

Naming and Program Execution at Present: Our Faculty, Our Students

This history has implications for the present and future success of the minor. In terms of our most immediate teaching needs, we continue to rely heavily on APLs, mostly extended-term, but occasionally—and more and more frequently in 4010—temporary lecturers. Again, concerns arise about "professionalism," as these temporary lecturers—although typically outstanding, if inexperienced teachers—often bring indirectly related or unrecognized professional and academic credentials to professional and technical writing instruction, in addition to having no job security and thus little time or incentive to seek additional credentialing, experience, or professional development. As a department, then, we need to think honestly and creatively about staffing solutions. And given the aforementioned uncertainty that many of those charged with hiring faculty in professional writing have brought to previous job searches, it seems clear that the challenge remains for our department to continue to raise the question of what "professional" means in our context—and how we will envision, staff, and deliver a "professional" course of study.

For our students, we need to think carefully about what our name communicates given the role it plays in marketing and recruiting. Who is our target audience? Does the minor deliver what it is supposed to? What do students think that is? One colleague notes reservations about the minor's suitability for English majors, a constituency we hope to reach:

> It's just that I think they're [courses in the minor] less useful to English majors than I had anticipated they might have been when we started. And maybe it's because the problems you have to address for the guy who wants to be a rancher who's trying to learn how to write reports are not the level of the person who wants to be the editor of a professional journal, for example.

One student, a graduated English major, shares this concern: "I think that sometimes, being already an English major, I felt that I wasn't getting quite as much from the required classes in the minor as other students might that are not already majoring in the field."

The same colleague suggests an important interpretation of our program name, clearly related to his belief about the minor's relevance to English majors: "The problem with it is that it's not professional writing. It's writing for people with different majors. People who are not going to become professional writers." In conducting our research, it seemed that at least some colleagues felt a similar sense of conflict and read or constructed the minor as a site where "professionalization" is the ultimate outcome. To put it in Couture and Rymer's terms, some faculty members expect the minor to graduate "career writers," rather than "professionals who write" (4-5). The distinction is meaningful here because it gets to the heart of the program's purpose and audience. If the minor is designed to develop "career writers," genuine writing professionals, it would seemingly exclude much of the external, interdisciplinary population of students who we believe 1) give it vitality and 2) stand to benefit from it. On the other hand, if the minor is designed and directed toward "professionals who write," some, at least, seem to see our own English majors as excluded. "Professional," here, is clearly ambiguous, and one can easily see the curricular complications that emerge. However, the ethical dimension of this confusion cannot be ignored: we must be certain that we are delivering the kind of instruction that benefits all of the students we invite; otherwise, we need to send out fewer invitations.

"Professional" Writing and the Challenge of Dissensus

We would like to make a few final observations about our minor's name, starting with the interesting paradox surrounding "professional": it evokes concerns about "practical," which is often seen as uniquely odious in English settings even as it evokes feelings of status—however authentic—among students, some faculty, and, we would guess, administrators. The implicit link between "practical" and a market economy can feel problematic to some, even as it excites others. One colleague, arguing for a more unified approach to literary and rhetorical education, notes about the minor:

> I think you've tried to subdivide or isolate a certain set of literary skills that don't necessarily depend upon a rich wealth of allusion or nuance.... if the extent of their power of allusion is the Microsoft homepage, then you're really working with a ceiling you'll never rise above. I don't think you can be really "professional" if your range of knowledge isn't beyond that kind of ceiling .... If all the allusions we make are to consumer values, then we're not advancing; we're not professional. We're not advancing the knowledge of our community.

To this colleague, achieving "professional" status means more than being practically prepared to join the workforce; indeed, "professionalism" is in some ways synonymous with liberal education and the intellectual discovery it implies. But again, other students, colleagues, and administrators may well hold a positive view of the practical dimension of "professional writing" (only one part of it, we would argue) as a necessary step toward a successful career and the status that follows. So, while "professional" is oft-hailed for its generality and the breadth it accommodates as a naming word, we must constantly be attuned to the very different ways in which students, faculty, and administrators will encounter it. We need to seek ways to unite the different goals these constituencies locate in the term "professional."

Optimistically, this variety of interpretations may well open the door to many possibilities for our program structure. However, this lack of consensus can lead to manipulation as well. Unlike composition and rhetoric or technical writing, for that matter, "professional" floats just above the surface of significance—it remains general and resists deep treatment of any *particular* kind of writing or communication. This absence of narrowly defined specialization can make it difficult, for instance, to argue for tenure-track faculty and extended-term APL lines, stalling deep investment in writing research.

More frustrating still is the fact that the term "professional" lets the minor be administratively manipulated, which may be simply unavoidable. Initial rumblings about a general "writing minor" turned at some point to a "professional" writing minor, which no doubt plays better at higher levels of administration even as its signification is relatively ambiguous. Again, this is to be expected to an extent. As one colleague notes, "I wouldn't expect there to be a real deep signification in this notion of professionalism, except that it's really kind of 'hitching on' to the cultural values of the early twenty-first century. Within the university we tend to live with an awful lot of euphemisms and packaging." But while this is true, it is frustrating that, for instance, connecting professional writing to the mission of the department's new MFA might be, as one colleague put it, "…largely rhetorical performance for the academic plan and academic affairs," if we are unwilling to work tirelessly to understand and professionalize the working conditions of those who teach in the minor that goes by the same name or fully acknowledge the intellectual depth of this emerging part of our English Studies "profession."

CONCLUSION

Names are important to program developers, but we believe they become even more crucial as a constituency gets farther away from the program. If we assume, for instance, that writing faculty and program developers enjoy the most intimate relationship with the program (perhaps a self-centered assumption, but one that makes intuitive sense) and, indeed, oftentimes are charged with the task of naming itself, other constituencies might be seen as existing—and thus *interpreting*—the program at some degree of removal. Other department members use their reading of the name, for instance, as an inroad into departmental conversations that can shape program direction and expectations. Similarly, students interested in enrolling in the program may rely disproportionately on the name to determine whether the program is relevant to their interests and career goals. In each case, these constituencies rely more heavily on the name for interpretation and decision-making than do those charged with developing the program. As the first interface many have with the program, a name matters.

Are the stakes lower for us at UW because we are talking here about a minor, not a major or graduate program? Perhaps. But we would argue that the core issues surrounding naming vis-à-vis program development and execution remain roughly the same. Any writing program must carefully consider the range of interpretations and expectations various constituencies will bring to bear on its curriculum and, significantly, its institutional role. Are these expectations at odds with one another? Which expectations can realistically be met given resources like faculty, funding, and goodwill?

Moreover, the ethics of recruiting students to a particular name must be of chief concern to *any* program. Students deserve to know what particular courses of study—minor, major, or graduate—can do for them in both their intellectual development and their preparation for a career. Naming—and operationalizing a chosen name—is central to this, as a program's label heavily conditions marketing and recruiting, as well as advising. Advisers, particularly those in other departments, frequently rely on either a limited understanding or a markedly different discourse when helping students make curricular decisions. These differences must be considered and accounted for. If signification must lack precision, writing program faculty members owe it to students and advisers to make as much information available as possible. We believe that examining that signification's history can go a long way toward developing a more robust shared understanding for all of these constituencies.

NOTES

[1] Kelly Belanger took a position as an associate professor at Virginia Tech University in fall 2005.

[2] In fall 1993, a change in status to extended-term Academic Professional Lecturers (APL) was implemented for qualifying temporary lecturers in the department. This position had a tenure track of sorts and lacked the previous term limits in keeping with the vision of the Wyoming Conference Resolution. For the English department, this change stabilized the ranks of instructors teaching heavily enrolled required composition courses. The original plan was to hire twelve APLs, four at a time over three years, but within two years of its inception, the dedicated funds for APL hires were rescinded and absorbed into the university's general fund; APL hires now competed with professorial hires for the same pot of money. Over the initial two-year hiring period, eight APLs were hired; only three members of that original group remain. Since then, four other APLs have been hired, with three still here, but these hires have been made at large intervals of time and often on a need-to-hire basis to fill critical gaps in other department offerings. One APL notes, "The English department has NEVER fulfilled, or even approached, its goal of twelve APLs."

WORKS CITED

Couture, Barbara and Jone Rymer. "Situational Exigence: Composing Processes on the Job by Writer's Role and Task Value" *Writing in the Workplace: New Research Perspectives.* Ed. Rachel Spilka. Carbondale, IL: Southern Illinois University Press, 1993. 4-20.

Cunningham, Donald H. and Jeanette G. Harris. "Undergraduate Technical and Professional Writing Programs: A Question of Status." *Journal of Technical Writing and Communication* 24.2 (1994): 127-37.

Dayton, David and Stephen A. Bernhardt. "Results of a Survey of ATTW Members, 2003." *Technical Communication Quarterly* 13.1 (2004): 13-44.

Faber, Brenton. "Professional Identities: What is Professional about Professional Communication?" *Journal of Business and Technical Communication* 16.3 (2002): 306-37.

Hayhoe, George F. et al. "The Evolution of Academic Programs in Technical Communication." *Technical Communication: Journal of the Society for Technical Communication* 41.1 (1994): 14-19.

Johnson, Robert R. "Plenary Address." Presented at the Thirtieth Annual Meeting of the Council for Programs in Technical and Scientific Communication. Potsdam, NY. October 2003.

Latterell, Catherine G. "Technical and Professional Communication Programs and the Small College Setting: Opportunities and Challenges." *Journal of Technical Writing and Communication* 33.4 (2003): 319-35.

MacNealy, Mary S. and Leon B. Heaton. "Can This Marriage Be Saved: Is an English Department a Good Home for Technical Communication?" *Journal of Technical Writing and Communication* 29.1 (1999): 41-64.

Mendelson, Michael. "Professional Communication and the Politics of English Studies." *WPA: Writing Program Administration* 17.3 (1994): 7-19.

Renz, Kathryn. "A Flare from the Margins: The Place of Professional Writing in English Departments." *Pedagogy* 1.1 (2001): 185-90.

Savage, Gerald J. "The Process and Prospects for Professionalizing Technical Communication." *Journal of Technical Writing and Communication* 29.4 (1999): 355-81.

Sides, Charles H. "Quo Vadis, Technical Communication?" *Journal of Technical Writing and Communication* 24.1 (1994): 1-6.

Sullivan, Patricia A. and James E. Porter. "Remapping Curricular Geography: Professional Writing and/in English." *Journal of Business and Technical Communication* 7.4 (1993): 389-422.

3 Composing a Proposal for a Professional / Technical Writing Program

W. Gary Griswold

The pages that follow describe the development of a grant proposal I wrote to obtain funding from an external private funding source. In the grant I proposed to research the programmatic options, student interest, and departmental/administrative support available for implementing a master's degree in professional and technical writing (PTW) at my university.

The lessons learned and offered here are many, but can be distilled thus: to put forward any type of proposal to develop (or as is the case here, to *investigate developing*) an academic endeavor as potentially complex as a graduate-level PTW program takes institutional savvy, administrative experience, and perseverance.

And even these can sometimes not be enough.

THE BEGINNING

It was just prior to my first semester as an assistant professor of English that, at a luncheon for new faculty, the Dean of the College of Liberal Arts mentioned to me that she had heard about a call for proposals that might be of interest. She knew of my experience teaching professional and technical writing, and this grant program provided start-up funds to work on establishing a master's degree in non-traditional fields.

If it sounds strange that a college dean would be chatting with a brand new faculty member about initiating an MA, it is important to mention that I was not really all that new. I had taught here at CSU (California State University, Long Beach), and run various English Department programs, including the writing center, for more than a decade, during which my contractual status had been short-term, temporary, or less-than-permanent; choose your favorite terminology. The previous semester I had been successful in my application for a tenure-track position, and I now found myself attending the same welcome and orientation meetings as new hires who had recently arrived from places like New Mexico, Arizona, and New York.

However, since I had been teaching at the campus for years and much of that within the technical writing certificate program we already had, I was no stranger to the idea of setting up a full PTW MA (or "PMA"). Over the years, the concept had been examined and discussed in various contexts, including meetings with the certificate program's advisory board, faculty teaching the classes, and various administrators. Ultimately, while most agreed that such a new program would be worthwhile, somehow no one had championed the cause, and it sat on a backburner for someday.

Certainly the grant award would give new life to this PMA concept, pushing it to happen sooner rather than later. However, it is likely that eventually we would have gotten around to doing much of what ended up in the proposal. I stress this point because it's relevant to a maxim a mentor of mine once told me: try not ever write a proposal for something you are not already doing, or at the very least, plan to do in the future.

Why? Well for one, (and here I speak mainly from my perspective as a faculty member in the CSU) you probably have a teaching and research load to keep you busy; anything else taken on should dovetail in some way for activities already underway. In addition, if you go about writing the proposal as I suggest here and consult widely with colleagues and administrators about it, you will likely get them excited about your idea. As a result, if it is not funded, you may well be encouraged to go ahead with your plan. It makes sense, then, to propose something that is doable within your regular workload, since your institution may or may not provide the same level of fiscal support asked of the external funding source. However, even if you propose something that seems like it will fit into present plans and work, the necessary tasks can well spiral out of control.

But I didn't ponder this very much prior to diving into this project; I just asked the Dean to forward me the grant information.

THE REQUEST FOR PROPOSALS

I tell my students that going after a grant is essentially detective work: the proposal writer, like a sleuth, has to look for clues. Rather than forensic evidence, though, the grant writer's job is to examine information concerning the values, preferences, interests, and taboos as expressed by the granting agency. Based on these clues, the grant writing gumshoe must consider whether or not the potential project is enough of a match (or can be made into a match) to the granting agency's agenda to warrant the necessary work needed to develop a good proposal document.

Therefore my first task was to read through the granting agency's Request for Proposals (the "RFP," as grant writers call it, the complete text of which is provided in Appendix A). The Council of Graduate Schools (CGS), the organization that had issued the RFP, had carried out some research on master's degrees for the Ford Foundation. CGS was now interested in receiving proposals for "planning grants," in which institutions would be given funding to support what were essentially feasibility studies, the purpose of which were to determine if student interest, employment demand, and institutional support existed for a "professional master's degree" in a given area of the humanities or social sciences. CGS hoped to hear about possible master's degree programs that "serve specific employment needs of business, industry, government, and non-profit sectors" (See Appendix A). They were particularly interested to fund grants describing activities that would include consultation with prospective students and area employers, as well as provide evidence (via letters from campus administrators at the department, college, and university levels) of institutional support for the plan.

All this had to be completed in a week and five pages.

Before doing anything else, I followed what I stress in my upper-division proposal writing seminar: I got on the phone. When most folks think of going after funding of some kind, they usually jump the gun and, after reading the RFP, devote a lot of time to writing. However, as is often taught in many a first-year composition course, a significant investment in what a writer does before beginning a draft will help ensure a better product. This guideline applies even more so in "real world" proposal writing, where it is possible to devote a week or more to writing and administrative scrambling only to find out that the proposal does not quite fit the funding agency's agenda.

So I called the director of this grant program at CGS to discuss my ideas and asked if my reading of their priorities was accurate. He, in turn, confirmed that I had a solid take concerning their RFP, but he also provided additional insight as well as motivation: CSG was particularly interested in programs that would result in master's degrees that were a professional end to themselves and not stepping stones to PhD programs or work inside academia. Also, though it was alluded to in the RFP, he confirmed that successful planning grant recipients would be invited to submit proposals for implementation grants, which would involve larger sums provided by the Ford Foundation, which was also providing the funding for these initial planning grants. He said a proposal relevant to a master's degree in technical and professional writing sounded very promising.

So at this point I had two very important "pre-writing" activities to engage in: garnering the necessary departmental "buy-in" for the idea (since I could not accomplish it on my own) and getting letters of support from administration. Also, I had this nagging but hazy feeling that I had probably better get going on the campus approval process as soon as possible.

SHOPPING THE IDEA

"It's a grant for a feasibility study. It's to see if we have student interest, faculty support, and the resources to develop an MA specifically in professional and technical writing."

"Ah."

"I've already seen a great deal of interest from many of our grad students. I suspect there's a considerable amount who don't want to be teachers or go onto PhD programs. They want to use their writing skills to get a job after the MA."

"Yes, well. That sounds great. I'm glad you're taking the lead on this. Maybe it will help get us other grants. Hey, have you considered designing a profit-making extension program? It could help us support our main MA."

Such was the reaction I got from one of the folks whose approval would probably be necessary to move beyond the feasibility study phase, should it ever come to that. I thought it best to let these people know about this "seed money" grant I was working on so that later on they would feel they had been consulted. So I had various informal chats with my fellow composition specialists.

What really surprised me is not that I was met with hostile stares, but rather *blank ones:* a couple of key players had a hard time seeing a PTW program as a legitimate—or even promising—segment of rhetoric and composition.

In "The Rise of Technical Writing Instruction in the America," Robert J. Connors concludes with a rather rosy view of the picture of the future of PTW programs, stating that "prospects have never been brighter," and that "it now seems likely that technical communication will be an acceptable field of study for English graduate degrees in many schools by the end of the decade" (96).

Well, twenty years later, at least in our department, that did not quite seem to be the case, but no matter: as the director of a writing center for over twelve years, I'd had a lot of experience explaining something outside the traditional academic scope to my colleagues. And so long as they did not seem eager to put up barriers, it seemed safe to proceed. However, I still needed to give a heads up to what I was doing to another member of our department.

As already mentioned briefly, our department already had a modest technical writing certificate program. For many years it had done well in serving students who wanted a small add-on to a traditional MA or BA in English. But for just as many years, various faculty who taught in the program as well as members of the program's advisory council had batted around the idea of having a full master's degree devoted to technical/professional communication. We suspected that student interest would be very high for such a program.

The problem was that no one seemed eager to take on the administrative nightmare that shepherding a new MA through the bureaucracy of the California State University system was said to entail. The director of the certificate program, a senior member of the department who had kept the program viable through the last decade's sundry fiscal crises, had often estimated that it would take at least five years to get a new MA degree approved. He certainly knew what he was talking about, but I also suspected that since he was nearing retirement, he did not want to initiate a new program that more junior faculty would have to deal with long after he was gone. This was also why, when I discussed the grant to him, he pretty much supported what I was doing, albeit with a knowing gleam in his eye. Since the grant was essentially to fund a "feasibility study," why shouldn't I take a look at the possibilities?

LETTERS, WE GET LETTERS

Assuming you are on good terms with your department chair and college dean (I was and thankfully still am), getting letters of support from them is fairly easy. However, obtaining such missives from provosts and presidents is another matter. Though they be wonderfully supportive, the problem remains that they often do not have any idea about your proposed program and probably do not have the time (especially when all you've got is a week) to sit down with you and discuss it.

Here's my method: First, I will usually find out who is the gatekeeper (or as a Hollywood celebrity's aide is dubbed, the "handler") for that administrator. This person is usually an "administrative assistant," though that title is not by any means universal. Next, after initially contacting the handler and determining that the administrator is willing to write a letter, I ask which the administrator would prefer: I can go ahead and draft the letter and send it to him/her to finalize, or I can send along information concerning the grant and then the administrator—or as is more likely, the administrative assistant—can write it.

All the while I am hoping that they choose the first option, since I do not really want the timely submission of my proposal resting on the hope that they will actually have the letter ready by the deadline. And I already have written the letter so it's immediately ready to email if and when they assent to the first option.

Last, there is the matter of (politely) making sure that the assistant and the administrator truly understand the time frame I'm working under. This is also a good opportunity to ask the handler to check the administrator's calendar to confirm that he or she will not be going out of town any time soon. I also emphasize I would love to stop by the administrator's office to pick up the letter as soon as it's done (and thereby not having to risk the vagaries of the campus mail system).

Thus, I am always ready to do a lot of walking the week the proposal is due and to grin and bear it after arriving and being told the letter is not quite ready.

CAMPUS CLEARANCE: THE PRODUCT GOES NOWHERE WITHOUT THIS PROCESS

Once again, the lesson to be learned is that having a great idea, the ability (other than funding) to carry it out, and the skill to write an effective proposal are not enough, especially if the deadline looms. If I had not been willing to literally walk the campus to hand-carry and personally pick up the various documentation necessary to the grant, the deadline would have been missed, even if I had twice the time.

A simple list of the non-writing tasks carried out relative to institutional clearance of my proposal will help illustrate:

1) Show the proposal to the College of Liberal Arts Development Officer and get her feedback.
2) At the urging of my Dean, make contact with the Director of Sponsored Projects to let her know the rather compact time frame I was working on so that I might get a quick lesson on the clearance process.
3) File a "Notice of Intent to Submit Proposal for External Funding." Make various phone calls to clarify how to do so.
4) Go to the Sponsored Projects Office to set up an "internal budget plan."
5) Make several phone calls to CGS and establish whether or not certain budget items would be allowed under the terms of the grant.
6) Provide a copy of the proposal text to various representatives of the Sponsored Project Office.
7) Initiate an Internal Clearance Form and confirm that it was making the rounds for the required approvals.

8) Meet in person with the Director of Research Compliance to make sure that my proposal did not have to be further examined by the Institutional Research Board.
9) Hand deliver the final draft of the proposal to the Director of Sponsored Research so that her assistant could mail it out (with a cover letter from their office) by the deadline.

Certainly all individuals involved did everything possible to make the process flow smoothly. However, a valuable lesson remains: with all the clearance steps required these days, a proposal writer *must* be willing to take a hands-on approach (or be fortunate enough to have a team of GAs or TAs to do so), or he or she may well endanger a great idea from being funded simply by missing a deadline. And even more important is the fact that the next time you need such clearance you do not want to be remembered as the person who made everybody scramble at the last minute.

Before moving on to examine the actual proposal, there are two tasks that were briefly listed above that deserve a bit more attention. The first is the somewhat cryptic (at least to first-time seekers of external funding) mention of an "internal budget." Very likely the money a proposal writer seeks will not in its entirety be applied to the proposed activities he/she seeks to fund; the institution must also cover its own costs. At my institution, all external monies awarded must be managed through an account at our Foundation Office. For every monetary transaction made, the Foundation Office charges a percentage to cover their costs for maintaining the account. Usually this "take" is only a modest amount of the total funding, but the point here is that when writing a proposal for external funding, it's usually wise to consider your budget page to be tentative until you have meet with your institution's folks responsible for overseeing external funding.

Of even more significance to anyone writing a similar proposal to garner funds for developing a program is what I breezily mentioned in item number eight above: getting your proposal cleared by your college or university's research compliance body.

A review of this kind is standard fare in the physical, applied, and social sciences, where faculty routinely go after hefty private dollars to fund research that in some way affects humans and/or animals in its investigation. However, for anyone whose primary academic activity is in the liberal arts/humanities, especially English, where many PTW programs reside, this whole research clearance concept may well be perplexing. Traditionally, "research" in our areas means, for the most part, examining artifacts (primary and secondary texts, etc.). Though we might assert these works "live and breathe," and some of the creators may still be alive, we are not doing anything physically invasive by writing about another's work.

It is when the proposal writer indicates that he or she will be sallying forth to talk, chat, interview, question, survey, or otherwise interact with actual people (either face-to-face or by other means) that we stray into the realm of using "human subjects," and at that point research clearance becomes an issue. I indeed proposed to interact with great many students, employers, and technical writing practitioners.

What saved me from having to seek formal approval and potentially months of delay (and thus missing the deadline) was the fact that what I was going to do was programmatic in nature. Here is a point where "administration" has positive connotations for the lone faculty member: since I was mainly going to consult with various interested people concerning the potential demand and interest in a PTW master's degree and use the resulting information for programmatic development (i.e., not publish it) I was in the clear. Such activity is considered within the scope of regular administrative duties for department chairs, program directors, and assistant professors looking to initiate new master's degrees.

WRITING THE PROPOSAL: SAYING WHAT YOU'LL DO

I do not intend to take you though each moment of my writing the week or so when I was actually drafting the proposal; going into what section I wrote first, how many drafts it took, etc. would not really be helpful here. However, I do think it would be useful to go through the sections of the proposal and discuss my rationale for exactly what I included. My main rationale for including each section, of course, was primarily that the RFP stipulated them in one form or another.

I started off the text of the proposal with a rhetorical device I encourage students to use in their professional documents: a concise purpose statement that informs the reader exactly what the document is all about. Then, the first paragraph of the "Rationale" section begins with a discussion of what I saw as the connection between the master's degree and the area employers. Here the proposal also stresses the importance of communication to the various enterprises mentioned in the RFP (business, industry, government, and non-profit sectors) while also briefly providing a definition of technical and professional writing and explaining its relevance to the Long Beach area in particular and the global marketplace in general.

Audience analysis, that concept we all come back to again and again with our students, is what was behind placing this information in such a prominent position; the RFP had stated that CGS was interested in funding programs that showed potential for meeting "local or regional workforce needs" (See Appendix A). Therefore, I wanted to establish here the employment and demographic diversity of the Long Beach area as I conveyed exactly what technical and professional writing was. In my experience, and the experience related to me by the practicing writers from the local chapters of the National Society for Technical Communication, most people, even those in academia (and even English Departments), have a very narrow view of what "technical/professional" writing entails.

In the next two paragraphs, the proposal begins to fully outline the "problem": few opportunities exist in California for the study of technical and professional writing, especially at an advanced level. Here some modest evidence is presented to suggest that such a demand exists generally, though specifically assessing the demand/interest more locally will be a part of the grant activities.

The last paragraph in the "Rationale" section overviews the problem that this proposal is intended to begin a process in solving. The intent here is to briefly demonstrate the kind of program that we would be looking to build.

The next section, "Relevant Institutional Background," was one that some campus colleagues and development experts I consulted with thought I might want to greatly shorten or leave out altogether since the proposal did not specifically call for it. Nonetheless, I still thought it was extremely important to include, though it did take away some space for other sections. Though some quick research on the CGS website indicated that their Board of Directors consisted of folks (mostly deans and provosts) who had extensive experience in higher education, I thought it best to provide some background on the CSU for three reasons: 1) even those experienced in higher education do not always know the particulars of other states' systems, 2) while the CGS Board consisted of people familiar with various higher education systems, the representatives of funding organizations they acknowledge in their literature, who I supposed would likely examine at least some of the proposals, would not be familiar with the CSU, 3) the RFP expressed interest in providing funding to institutions with "a track record of admitting students to master's degree programs ... rather than offering master's degrees only to students admitted to doctoral programs, but who do not complete the doctoral degree" (See Appendix A). I dang well wanted to be sure that they knew that this very much characterized CSU Long Beach.

The next section, "Proposed Plan of Action," as its name implies, outlined what I said we would do with the funding if awarded. Using the exact wording in the proposal, these are listed below:

1) Gather quantitative and qualitative data regarding student interest and perceptions for a master's degree in technical and professional writing.
2) Consult with Employers and Current Professionals.
3) Fully assess the feasibility and options of establishing the master's degree program.
4) Draft an action plan for implementing the master's degree program in technical and professional writing.

There is nothing particularly innovative about the action areas listed here; they are appropriate steps for assessing the feasibility of any academic program, and all are in line with what the RFP outlined as fundable activities.

HOW THINGS STAND

And funded they were: the proposal was submitted on time, and a few months later I learned that CGS had awarded us the grant. For a good part of a semester and much of the summer break, I have been carrying out what was promised in the proposal: wide consultation with students, faculty, technical writing practitioners, and employers in order to establish the need for the MA and begin considering the shape it might take. For me, as a "new" faculty member, I have been able to make invaluable contacts both on our campus and in the surrounding business community and have been amazed at how all sorts of folks will pay attention when I mention that our project is "supported by funds from the Ford Foundation." I also now have a thorough understanding of the campus clearance process for faculty going after external funding.

Which is a good thing, since the process is about to start all over again. CGS has invited us to submit an implementation proposal: a report of the grant's activities thus far as well as a detailed description of how we would like to set up the program. In fact, as soon as I complete the manuscript of this very article and email it to the editors of this volume, I plan to jump upon drafting that proposal and initiating the clearance and support-letter gathering processes. I am glad to say that I have allowed a bit more time to finish this one; the deadline is once month hence, but, of course, the proposal is lengthier and the required letters of support both more numerous and specific.

The only glitch is that CGS has yet to receive final confirmation from the Ford Foundation that it will provide the considerably larger funding for the implementation grants. CGS has nonetheless encouraged us to submit these proposals for follow-up grants.

AN UPDATE

It has been just about a year since I completed the terms of the planning grant. As outlined in the proposal, I held meetings with local PTW practitioners as well as those who employ them. I also conducted a survey of current English majors at CSULB and orchestrated a focus group discussion of several students enrolled in our upper-division PTW courses. In brief, these meetings established that there did indeed exist a great deal of interest in this program.

As another aspect of the grant, I consulted far and wide on our campus concerning what it would take to set up a new PTW master's degree in our department. What I discovered was quite close to what the director of our certificate program had told me: it would take around five or six years at the very least to get approval from the required department, college, university, CSU system, and California bureaucracies, and that did not include developing the curriculum.

Nonetheless, I did discover a much more expedient way to essentially accomplish the same thing: we could develop an area of emphasis within our current MA. We already had ten such emphases, ranging from medieval literature to rhetoric and composition. The only thing was that just as with all those other emphases, the actual degree would say "Master of Arts in English" with no mention of the particular concentration. However, I and other faculty from the department didn't think that would be too great an impediment since it had never dissuaded students from enrolling in the others.

So, with input from a couple of interested colleagues, I wrote up a rationale describing the need and interest in and for the new MA emphasis and a description of its curriculum. This included courses that we already offered with a few more that would be developed (and that I and the others had wanted to put together for quite some time). Ultimately, the document I developed became the proposal for an "implementation grant" that CGS had invited us to submit, whereupon I spiraled back into the previously described support-letter-gathering and clearance-obtaining maelstrom.

Unfortunately, we were not awarded an implementation grant. My follow-up inquiries confirmed what I suspected were the reasons: our proposal was not a bad one, but competition was extremely fierce for less funding than had been expected. As a result, very few institutions received the second level of grant funding.

But that wasn't quite the end of the PTW master's degree concept. I eventually had to report to my Dean that we didn't get implementation funding, but she was still enthusiastic and encouraged me (as did several other senior faculty) to pursue the idea. It was also hinted that on-campus funding might be available to provide some support. At any case I was congratulated for what I had accomplished so far.

Excited, I shared this information with the English Department Chair. She also congratulated me on what I had done and urged me not to fret about not getting the second proposal. However, she surprised me by expressing a strong disinclination that I continue with the project. Below are the points she raised.

At our institution at least, it seemed that getting a grant proposal funded and carrying out its terms is an activity that falls into an indeterminate grey area for the purposes of retention, tenure, and promotion, the trinity of work that no assistant professor can afford to ignore. Is writing a funded proposal equal to a publication? To publication in a juried journal? Or is it just service? In the case of someone like me, who teaches upper-division seminars devoted to things like proposal writing, does it count as teaching development? Regardless of the category writing a successful grant proposal falls into, and using the parlance of my institution, another important question was did such work meet "essential" or merely "enhancing."

The written policies offered little clarity. While the Retention, Tenure, and Promotion policy documents of the College of Liberal Arts hinted that funded proposals could be considered "essential" items of scholarly research, our department policies pretty clearly categorize them as "enhancing" (i.e., "lesser"). Though I had done well in my three-year retention review, it was clear that there would be some doubt as to whether all this grant seeking was to count for much in another couple of years when I was up for tenure and promotion.

My Chair, rightly so, was concerned that continuing with the PTW master's degree would get in the way of my publishing scholarly work. The former endeavor was something of a crapshoot concerning how it would contribute to my six-year review; the latter was certain to be seen at "essential." As a result then, of its nebulous scholarly status, I was strongly advised to shelve the PTW master's idea for the time being. To return to the detective metaphor raised at the outset of this article, I was taken off the case.

The proposal outlining the option in professional and technical writing for our current master's program remains on my office computer and archived on my back-up flash-drive. I've shown a printed copy to a few of my newly tenured colleagues and halfheartedly suggested they begin shepherding it through our department and college curriculum process—lest the momentum gained for the idea is lost during the next three years.

So far, no one has taken up the offer.

APPENDIX A: THE ORIGINAL RFP

Council of Graduate Schools
Professional Master's Program in the Social Sciences and Humanities
Request for Proposal
July 28, 2003

The Council of Graduate Schools [CSG] invites proposals for planning grants to support the needs assessment and development of professional master's degree programs in the social science and humanities fields (PSSHM). Professional master's degree programs prepare graduates for non-academic employment that serves local or regional workforce needs rather than for doctoral study.

For the past two years, CGS has supported the development of professional master's programs in science and mathematics fields (see www.sciencemasters.com). With support from the Ford Foundation, CGS recently conducted a survey of master's education in the social sciences that generated interest among social science and humanities disciplinary societies for a collaborative research and demonstration project that assesses the need for and promising models of professional master's programs. The CGS/Ford program will provide grants to CGS members to participate in this initiative.

Eligibility

All CGS member institutions that meet the following criteria are eligible to participate in this program:
- The institution must have a track record of admitting students to master's degree programs in the general disciplines specified for the project, rather than offering master's degrees only to students admitted to doctoral programs, but who do not complete the doctoral degree.
- The institution and participating departments must have adopted strategic goals that are consistent with developing PSSHM programs that respond to non-academic employment needs. Letters of endorsement from department chairs and administrative officials (including the graduate dean and chief academic officer) can be appended as evidence of commitment of faculty effort and institutional resources to the proposed PSSHM planning grant process.

Summary and Scope

The CGS/Ford project will provide grants to a significant number of member universities to participate in the collaborative research and demonstration project on professional master's education. The core of the project is the development of models of professional master's programs that serve specific employment needs of business, industry, government, and non-profit sectors. The models will provide additional insight into the trajectories of master's education in relation to societal needs. We are hopeful that the results of this PSSHM planning/development project will provide a compelling basis for a second series of grants to implement some of the proposed PSSHM programs.

CGS will make a maximum of 60 PSSHM planning grants of up to $6,000 across a broad range of social science and humanities disciplines in universities reflecting the variety of CGS member institutions: private and public; minority and majority serving; research/doctoral and master's focused. This coverage will demonstrate most effectively the broad applicability of the concept of professional master's education and provide sufficient numbers of models to attract the attention of colleagues and peers and to serve as templates for replication of the programs in other departments and institutions.

Activities to be undertaken in the assessment and program planning process
The grants will provide support for activities such as:
- Contacting prospective non-academic employers and engaging them in a discussion with departmental faculty and institutional officers concerning the skills and backgrounds they expect of new employees and realistic projections of workforce needs for PSSHM graduates.
- Establishing an external board to advise on curricular issues, offer information, serve as external mentors to PSSHM students, and sponsor internships for students in PSSHM programs.
- Conducting information sessions/focus groups/surveys among likely pools of prospective PSSHM students in order to determine interest and to project enrollments.
- Assessing institutional and departmental commitments to and capabilities of developing PSSHM programs, either by establishing new degree programs or by revising existing master's degree programs and incorporating professional components.
- Developing a proposal for implementing one or more model PSSHM programs, provided employer, student, faculty, and institutional support are sufficiently strong. The proposal will include a PSSHM curriculum with appropriate disciplinary core strength, components that develop high-level communications and professional skills, and employer commitments for internship experiences. A business plan will be required that includes projection of tuition appropriate for the applicant pools, contributions for program funding from employers

and the institution, internship stipends/salaries, and other revenue sources that allow the program to be developed and sustained at a cost acceptable to the institution.

Application materials for an assessment and planning grant
Institutions are encouraged to submit proposals for as many as three PSSHM assessment and planning grants per institution. The body of the proposals for each program area should not exceed five pages[2].

The proposals must:
- Demonstrate an interest in and commitment to master's education by faculty and the institution, including appropriate credit for faculty through the university review and reward system
- Propose strategies to seek the participation of minorities and other underrepresented groups
- Commit appropriate matching funds and effort to accomplish the goals of the planning grant: in most cases we anticipate these goals would include a proposal to create a PSSHM program
- Indicate an interest in establishing PSSHM programs in two or more departments
- List activities to be used to determine needs, interest, and institutional capacity for developing PSSHM programs

Appended material as required to:
- Document that faculty from departments that would be most likely to develop PSSHM proposals are committed to the project (department letters that express interest in and commit faculty efforts to the project are particularly relevant.)
- Assure that the activities and intent of the grant are consistent with and complementary to the institutional mission and strategic plans (a letter of endorsement by the chief academic officer or president would be particularly useful.)
- Provide evidence of endorsement by the graduate school (a letter from the graduate dean or other person responsible for graduate education at the institution.)

Project time-line
July 2003: CGS sends RFP to member institutions and posts on CGS website
October 15, 2003: Deadline for response to RFP for CGS/Ford PSSHM planning grant
Oct.-Nov. 2003: Evaluation of proposals in response to RFP
November 2003: CGS/Ford PSSHM grants awarded to graduate deans
December 2003: CGS Annual meeting: plenary session on professional master's education and meeting/progress reports for deans, directors of CGS/Ford PSSHM projects
May 2004: Interim reports due from PSSHM planning project directors/graduate deans

July 2004: Meeting of PSSHM deans and directors at CGS Summer Workshop
September 2004: Final reports due for CGS/Ford PSSHM planning grants. Proposals for implementing proposed PSSHM programs due

Responses to this RFP in the form of proposals for a CGS/Ford PSSHM Planning Grant may be sent via e-mail (preferred) or by U.S. mail (with an e-mail notice that proposal is being sent).

Send completed applications to For more information or questions, contact
Council of Graduate Schools Les Sims or Peter Syverson
Professional Master's Degrees lsims@cgs.nche.edu or psyverson@cgs.nche.edu
One Dupont Circle, NW, Suite 430 Phone: (202) 223-3791
Washington DC 20036 FAX: (202) 331-7157
www.cgsnet.org

[1] Professional Master's Programs in the Social Sciences: Current Status and Future Possibilities, Report to the Ford Foundation. The Council of Graduate Schools, Washington, DC 2003. Available upon request.
[2] The Professional Science Master's startup checklist (http://www.sciencemasters.com/startup_checklist.html) provides a set of topics that could be useful in developing a PSSHM proposal.

APPENDIX B: THE PROPOSAL
A Professional Master's Degree Program in Technical/Professional Writing:
A Planning Grant Proposal

Prepared for:
The Council of Graduate Schools
Professional Master's Program in the Social Sciences and Humanities
Prepared by:

W. Gary Griswold, PhD
Department of English
California State University, Long Beach
1250 Bellflower Blvd.
Long Beach, CA 90840-2403

Purpose Statement
This proposal outlines a four-part process to enable faculty at California State University, Long Beach (CSULB) to develop an implementation plan for a Professional/Technical Writing program leading to a Master of Arts degree.

Rationale
The city of Long Beach, California, its surrounding communities, and the greater Southern California area abound with corporations, government agencies, and non-profit organizations requiring the advanced skills of specialized communicators. Simply stated, prose communication is a staple of success; no organization in this region can thrive and grow without professional writers. In all forms of print and electronic media, the technical and professional writer in Southern California is the conduit to a wide variety of internal and external audiences with diverse linguistic, cultural, and demographic profiles. The scope of this diversity becomes evident with the realization that over 40 different languages and dialects are spoken in the city of Long Beach (LBUSD, 2003).

However, few opportunities exist in California for graduate-level study in the area of technical and professional writing. The Society for Technical Communication's national Academic Programs Database lists only 15 universities in California offering **any** coursework in technical and professional communication. Only five of those are within a 50 mile radius of the Long Beach area, and all of these are certificate and/or extension programs. In fact, there exists no master's degree program whatsoever in California focusing specifically on technical and professional writing (STC, 2003).

Nonetheless, the websites of the Los Angeles and Orange County chapters of the Society of Technical Communication list more than 50 current job announcements in the area (LASTC, 2003; OCSTC, 2003), and the U.S. Department of Labor's *Occupational Outlook Handbook* indicates that the employment rate of such positions "is expected to increase faster than average" (U.S. Department of Labor, 2003). *Money* magazine has suggested that work as a communicator in scientific, technical, medical, and other specialized areas is one of the top twenty best careers in the nation (Gilbert, 1994).

A professional master's degree program in technical and professional writing will provide students the opportunity for in-depth study of the advanced rhetorical and compositional theories and practices necessary for them to be leaders in designing, composing, and editing the prose and visual media so critical to success in nearly every industrial, scientific, governmental, technical, corporate, institutional, and philanthropic endeavor. This same master's degree would offer academic study and training concerning communicating with multicultural and multilingual audiences. In addition, completion of such a program would confer upon students the appropriate advanced credential often required by employers for career advancement.

Relevant Institutional Background
The California State University (CSU) system consists of 23 campuses offering bachelor's and master's degrees in more than 1,600 programs in approximately 240 subjects. (CSU, 2002). CSULB is the largest campus in the California State University system and the second largest institution of higher education in the state.

Currently, the CSULB Department of English offers a Technical and Professional Writing (TPW) Certificate Program. Earned in conjunction with a baccalaureate or master's degree (usually, but not always in English or related areas), the program requires 24 units of coursework. A capstone portfolio project and internship experience are also required. A small Advisory Council, consisting of various area practitioners (in both freelance and in full-time positions), assists faculty in the overall direction of the program. As valuable and established as the TPW Certificate Program is, in its current form it cannot offer the scope and contemporary focus that a full master's degree program would provide.

The vision set for here, then, is to allow our faculty to build upon the foundation of the current TPW Program at CSULB, so that it may develop into the leading technical and professional writing graduate program it has the potential to be.

Proposed Plan of Action
The following four action areas are proposed as most critical in developing a sound academic and financial plan for a master's degree in technical and professional writing at CSULB. Under each are listed the proposed activities to be carried out with the support of this planning grant.

1) Gather quantitative and qualitative data regarding student interest and perceptions for a master's degree in technical and professional writing: Methods to be used include written surveys as well as focus groups. Current graduate students as well as graduating baccalaureate students will be included in these efforts. Expertise from those faculty and staff involved in the Department of Communication's Hauth Center for Communication Skills can be drawn upon for establishing and conducting the focus group interviews.

2) Consult with Employers and Current Professionals: Area employers as well as practicing technical communicators will be consulted in a needs assessment process concerning the advanced skills and knowledge they see as important to include in such a master's degree. Special emphasis will be placed on involving, whenever possible, those practitioners and employers from minorities and other underrepresented groups. The TPW Certificate Advisory Committee will be a valuable resource in this activity, which will likely result in a wider range of interested employers and practitioners participating in the advisory board established for the proposed master's degree program.

3) Fully assess the feasibility and options of establishing the master's degree program: An interdisciplinary and interdepartmental range of faculty, staff, and administrators will be consulted concerning the options available for establishing the master's degree.

Faculty, staff, and administrator expertise can be drawn upon from a number of departments and programs both within and external to the College of Liberal Arts: the Department of English's Composition Program, the Department of Communication Studies, the Department of Computer Science, the Department of Political Science, the Center for Language Minority Education and Research, the Hauth Center for Communication Skills, the Writer's Resource Lab, the College of Engineering, the Department of Journalism, the College of Business Administration, and the College of Natural Sciences and Mathematics.

4) Draft an action plan for implementing the master's degree program in technical and professional writing. Led by the principal investigator, faculty and staff involved in the planning process will draft an action plan for the technical and professional writing master's program. Drawing upon the input gathered in the consultation process detailed above, this document will outline the most feasible program structure as well as estimates of costs and available funding.

Throughout the planning, faculty and staff involved with remain cognizant of the fiscal climate likely to continue for some time in California. Plans for having the program generate, whenever possible, its own revenue streams and strategic development plans will be examined along with traditional programmatic funding sources. In addition, throughout the planning process the integration of outreach to minority/underrepresented student populations will be a priority.

Course release time will be provided to the Project Director by the Department of English and the College of Liberal Arts to facilitate this planning grant. (A normal workload at this campus is four 3-unit courses per semester.) During the preparation of the action plan as well as any ensuing proposal development for funding and implementation, other faculty will participate as time and resources are available.

It is important to note that the Department of English has the expertise necessary to develop and carry forward the planning process here, as well as the implement the resultant master's degree program. In the last three years alone, three tenure-track faculty have been hired with experience in general technical writing and rhetoric, visual literacy, new media, applied writing technologies, and scientific/technical editing. In addition, a core of five full-time lecturers regularly teach classes within the current TPW Certificate program. Among these eight faculty, six bring extensive marketplace experience in technical and professional writing.

Conclusion
CSULB is uniquely positioned to develop a viable professional master's program in technical and professional writing, one that can draw upon its broad array of institutional expertise and resources to provide students with a learning experience that is grounded in current theoretical and practical applications. It is hoped that once such a program is established, the planning process and the actual program may serve as models for other institutions to develop their own professional master's degree programs.

References
Bureau of Labor Statistics, U.S. Department of Labor, *Occupational Outlook Handbook, 2002-03 Edition,* Writers and Editors. Accessed online at <http://www.bls.gov/oco/ocos089.htm>. (Date of access: October 4, 2003).
California State University, "About the CSU." Accessed online at <www.calstate.edu>. (Date of access: February 13, 2002)
Gilbert, J. "The Best Jobs in America." *Money.* 23.3 (1994): 70-73.
Long Beach Unified School District, "About the District." Accessed online at <http://www.lbusd.k12.ca.us/research/update2001/index.html>. (Date of access: October 7, 2003).
Los Angeles Society for Technical Communication. Accessed online at <http://www.lastc.org> (Date of access: October 1, 2003).
Orange County Society for Technical Communication. Accessed online at <http://www.stc.org/academic.asp> (Date of access: October 1, 2003).
San Gabriel Valley Society for Technical Communication. Accessed online at <http://www.stcsgv.org> (Date of access: October 1, 2003).
Society for Technical Communication, "Academic Programs." Accessed online at <http://www.stc.org/academic.asp> (Date of access: October 1, 2003)

Tentative Schedule

December 2003/January 2004:
Plan, organize; and design survey instruments and focus group sessions; attend CGS Annual meeting (Dec. 3-6), assemble core faculty team; hire GA assistance.

February/March 2004:
Administer surveys; conduct focus groups; consult with faculty, staff, and administration in other departments; develop employer and practitioner contacts.

April/May 2004:
Analyze survey response, focus group, and interview data; begin drafting of action plan; circulate action plan draft for institutional input.

June/July 2004:
Prepare final draft of action plan; attend Summer Workshop (July 10-14).

WORKS CITED

Conners, Robert J. "The Rise of Technical Writing Instruction in America." *Technical Writing and Communication.* 12.4 (1982): 329-51. Rpt. in *Teaching Technical Communication: Critical Issues for the Classroom.* Ed. James M. Dubinsky. Boston: Bedford/St. Martin's, 2004. 77-98.

Council of Graduate Schools. *The Council of Graduate Schools: Advocacy, Research Innovation.* Washington, D.C.: Council of Graduate Schools, 2004. 10 September, 2004. http://www.cgsnet.org/aboutcgs/brochure.htm

Gage, John T. "On 'Rhetoric' and 'Composition'." *An Introduction to Compositions Studies.* Ed. Erika Lindemann and Gary Tate. New York: Oxford UP, 1991. 15-32.

4 Disciplinary Identities: Professional Writing, Rhetorical Studies, and Rethinking "English"

Brent Henze
Wendy Sharer
Janice Tovey

> *More than a program's course content, a curriculum is a contested representation of the public identity of an institution and a discipline.*
>
> – David B. Downing, Claude Mark Hurlbert, and Paula Mathieu (1)

Few of us in professional and technical writing or rhetoric and composition have avoided the turf wars that often accompany the development or revision of curricula within English departments. At institutions across the country, faculty in these areas have attempted to carve a niche for themselves, often in the midst of heated resistance. When we, along with several of our tenured and untenured colleagues in the areas of rhetoric, linguistics, and professional writing, proposed a curriculum for an undergraduate concentration in "Rhetorical Studies and Professional Writing" (RSPW) as one option for English majors at East Carolina University (a regional state university with approximately four thousand graduate and sixteen thousand undergraduate students), we certainly felt some heat.

At the time we developed our proposal, the department offered two degree options: 1) a BA in English, and 2) a BA in English with a Concentration in Writing. The former option provided students with detailed knowledge of literary periods and genres, while the latter combined course options in creative writing, technical and professional writing, and rhetoric and composition. As we describe below, faculty specialists in technical and professional writing and rhetoric and composition, frustrated by curricular limitations within the Concentration in Writing, developed a new curriculum that would provide students with an opportunity for focused study of writing in a variety of professional and civic contexts. While the Concentration in Writing had made sense when it was developed—a time in which creative writing courses constituted the majority of writing courses in the department—the growth of course offerings and faculty presence in rhetoric and professional writing seemed to call for curricular revision. We envisioned that our proposal would lead to two separate, but ultimately more purposeful, concentrations in writing—one a rhetorically based study of writing in the workplace and the community, and the other a creative writing track with course parameters and requirements to be developed by creative writing faculty. We intended for the RSPW concentration to provide both theory and practice for undergraduates interested in a variety of writing-related careers. In order to provide this focused study, our revised concentration omitted a few of the previously required literary history courses required of all students in the existing BA curriculum. While we suspected that these changes might meet with some resistance, we hoped that, by retaining some of the literature requirements and by making the argument that this new concentration would increase the total number of English majors, we could persuade our colleagues that the revisions were in the department's best interest.

At a rather contentious faculty meeting, the motion we put forward to incorporate the new RSPW concentration was critiqued and ultimately tabled pending further discussion among department faculty. The attempt to reconfigure our offerings to English majors met with significant resistance and exposed tenuous relationships among the disparate scholarly and pedagogical interests in our department. We had thrown open the floodgates of disagreement about what a degree in "English" means. The proposed changes prompted faculty in the department to engage in heated, sometimes painful, but ultimately necessary conversations about what the "core" courses in the department should be and what a "core" in an English Department should accomplish.

We discuss three aspects of our proposed program here: 1) The structure and rationale of the revised curriculum; 2) The departmental identity issues our proposal raised, including the instability of disciplinary boundaries that demarcated the department's programs in the past; and 3) The tactical changes we would make if we could start this process anew. This article is not intended to be a gripe session—such an indulgence would assist neither us nor our readers. Instead, we discuss our conflicts, frustrations, and missteps in the spirit of working through them. How might the process of proposing a new curriculum have been better executed? What problematic assumptions and communicative practices impeded our attempts to revise the curriculum, and what might we have done to better respond to these problems? We conclude our discussion on a positive note, with a brief overview of some of the positive results that this struggle has produced. We believe that our story, through the cautionary tales and advice it provides, will interest other faculty and administrators who are just embarking on the process of constructing an undergraduate program in technical and professional writing, and who are, in the process, redefining the boundaries among traditional and emerging specializations within English departments.

PROPOSED REVISIONS

At the time of our proposal, the department awarded two Bachelor degrees—a BA in English, which focused primarily on literature; and a BA in English with a Concentration in Writing. In the past, the department has offered a BA in English Education, but that program was recently relocated to the College of Education. The BA in English with a Concentration in Writing was first listed in the 1978-79 undergraduate catalog, and, although two distinct BA concentrations were elaborated beginning with that catalogue, all English majors began their undergraduate major programs with a "common core" of courses. The nature of the courses included in this common core has changed over the years, with recent configurations of the core requiring students to take a fairly specific sequence of literary studies courses. The chart below shows the 2004-05 core:

2004-05 Undergraduate Catalogue

Core Courses (required of students in both Concentrations)
- ENGL 2000. Interpreting Literature
- ENGL 3000. Lit in English to 1700
- ENGL 3010. Lit in English, 1700-1880
- ENGL 3020. Lit in English, 1880-Present
- One Shakespeare course (Tragedies, Comedies, or Histories)

For English (Literature) Concentration
- One course in language or composition (includes courses in linguistics, composition and rhetoric, and creative writing, but not professional writing)

For Writing Concentration
- One non-writing elective; choices include linguistics, film studies, as well as other literature courses

---

FIGURE 1: CORE REQUIREMENTS, 2004-05 CATALOGUE

Prior to our proposal for an RSPW curriculum, the consensus among a significant segment of the faculty for several semesters had been that we needed to reevaluate and reconfigure our undergraduate major and consider possible changes, particularly to the overburdened and perhaps overly specific core. The reasons for such a reconfiguration were several:

1. **Use of faculty strengths:** The course options in the existing curriculum did not take advantage of the department's growing number of faculty (fifteen at the time of this writing) with strengths in technical and professional writing, discourse analysis, linguistics, and rhetoric and composition, particularly at the upper-division undergraduate level. For example, faculty specialists in rhetoric and composition had only one upper-division course at the undergraduate level: English 3810: Advanced Composition.

2. **Recognition of disciplinary diversity:** The limited selection of core and elective courses in these areas reflected a lack of recognition and value for these areas in the undergraduate curriculum.

3. **Response to student course needs:** The selection of writing courses was heavily weighted toward creative writing and did not address the needs of students in the writing concentration who were not interested in creative writing. Creative writing workshops—two each in poetry, fiction, playwriting, and creative nonfiction—offered a variety of writing experiences for those interested, but there was a noticeable absence of courses in rhetorical theory and composition studies. Other "non-creative" writing courses included two services courses, business writing and scientific writing (both grandfathered in when the WAC program established writing requirements in all disciplines), and only three courses created especially for professional writing: editing, publications development, and internships.

4. **Coordination of departmental programs:** The undergraduate curriculum was not clearly coordinated with the graduate curriculum. We believed that our strong MA program in professional and technical communication would be enhanced by a strong BA in writing, as would the department's PhD program in Technical and Professional Discourse. This PhD program includes three focus areas: technical and professional communication, writing studies and pedagogy, and discourses and cultures. While the third area—discourses and cultures—was represented in several of the department's literature courses, the first two areas and the study of linguistics (an important component of the third area) lacked a strong emphasis in the undergraduate major.

In response to these situations, we proposed a curriculum that decreased the number of required (core) courses, created new courses in both rhetorical studies and professional writing, and provided more flexibility for students:

2004-2005 BA IN ENGLISH (36 S.H.)	
Concentration in English	Concentration in Writing
2000. Interpreting Literature	2000. Interpreting Literature
3000. Lit in English to 1700	3000. Lit in English to 1700
3010. Lit in English, 1700-1880	3010. Lit in English, 1700-1880
3020. Lit in English, 1880-Present	3020. Lit in English, 1880-Present
One SHAKESPEARE (Comedies, Histories, or Tragedies)	One SHAKESPEARE (Comedies, Histories, or Tragedies)
One course in LANGUAGE or COMPOSITION (includes courses in linguistics, composition, and creative writing, but not prof comm)	Six courses in WRITING (includes courses in composition, prof comm, and creative writing)
Six ENGLISH ELECTIVES (excludes writing courses)	One ENGLISH ELECTIVE (excludes writing courses)

FIGURE 2A: COMPARISON OF 2004-2005 AND PROPOSED BACHELOR OF ARTS CURRICULA IN ENGLISH

PROPOSED BA IN ENGLISH (36 S.H.)		
Concentration in English Literature	Concentration in Rhetorical Studies and Professional Writing (RSPW)	Concentration in Creative Writing
2000. Interpreting Literature 3000. Lit in English to 1700 3010. Lit in English, 1700-1880 3020. Lit in English, 1880-Present	2000. Interpreting Literature One upper-level lit course One course from linguistics, film, folklore or creative writing	To be defined by faculty in creative writing
One SHAKESPEARE (Comedies, Histories, or Tragedies)	3030. Intro to Rhet Studies* 3040. Intro to Prof Wtg* 4885. Capstone Semr in RSPW*	
One course in LANGUAGE or COMPOSITION (includes courses in linguistics, composition, and creative writing, but not prof comm)	Four courses in RSPW, with at least one from each area: A. Professional Writing B. Rhetorical Studies	
Six ENGLISH ELECTIVES (excludes writing courses)	Two ENGLISH ELECTIVES	

* indicates proposed new course

FIGURE 2B: COMPARISON OF 2004-2005 AND PROPOSED BACHELOR OF ARTS CURRICULA IN ENGLISH

Once our first draft was submitted informally to the faculty for review, the department held two open meetings for faculty to respond to the proposal.

These meetings highlighted conflicts that, although present for many years, had been generally overcome by a collegial environment. In a department in which rhetoric and professional writing courses are strong programs at the MA level (the MA degree in Technical and Professional Writing, for example, has a maturing online degree with about seventy-five active students, a majority of the department's MA students) and are a significant factor in a PhD program, some of the faculty not specifically involved with these areas became concerned and disturbed by our proposed reduction in the number of required literature courses for students pursuing the RSPW concentration. Some even suggested that since RSPW faculty "had [the] PhD," we should let literature faculty "have the BA." Our attempt to extend our presence to the undergraduate major became a territorial struggle. The extent to which reactions to our proposal reflected struggles over institutional power can be seen in the responses of faculty in other areas of study that were also not firmly rooted in the departmental core. While we were perceived by some faculty as upstarts trying to deny the "soul" of English—the study and composition of literature—our proposal garnered support from faculty teaching in other areas with limited visibility in the major: linguistics and multicultural literature.

THE MEANING(S) AND BOUNDARIES OF THE "ENGLISH" MAJOR

Our conclusion that the resistance that we encountered was tied up in larger territorial battles is hardly earth-shattering. It will be more revealing, and, we hope, helpful to our readers for us to analyze the specific scholarly and institutional conditions that demarcated disciplinary boundaries and subsequently fueled that resistance. On the one hand, these sorts of conflict may be the inevitable consequence of gathering together any group of twenty or fifty or (in our case) eighty strong-willed people who are devoted to what they do. On the other hand, while there is no doubt that ego and personality conflicts have had some part in this resistance, we do not want to reduce the divergence of opinions to "interpersonal conflict" or to present that divergence as simplistic "us vs. them" factioning. Rather, we would like to outline some of the differing assumptions that inform the department's guiding terms and that maintain (or challenge) the boundaries among the many scholarly pursuits currently housed within the English department. Departments of all shapes and sizes have had to deal with just the kinds of conflict that we have wrestled with: conflicts over the real "meaning" of the English major, the role of the so-called "practical" courses in an English curriculum, and the merits and problems of institutionally separating writing from literature.

In his concept of "disciplinary boundary-work," sociologist Thomas Gieryn offers a useful lens through which to examine the controversies that arose within our department (see also David Russell's discussion of boundary work in the composition/literature split). According to Gieryn, a discipline's representatives strategically shape its boundaries by means of discourse: they articulate the discipline's mission in a certain way, they define a set of characteristic problems to coincide with the discipline's methodologies, they articulate collective values, and they engage in other practices to widen the discipline's scope and strengthen its resources. In Gieryn's approach, the epistemological, ontological, and practical relationship between a discipline and the surrounding culture is interpreted according to a cartographic metaphor. Gieryn employs this familiar metaphor to explain that a discipline relates to other disciplines, and to larger systems of knowledge and activity, in the same manner as a geographic territory relates to neighboring territories and to the larger land mass that encloses it. Furthermore, the relationships between neighboring territories strongly influence the overall health, power, and legitimacy of the involved territories. As such, it is helpful to know how the boundaries between territories are formulated and how they share resources.

What's up for grabs in boundary conflicts is not just traditional "resources" (such as faculty lines, research funds, courses, and students), but also control over representations of the discipline's central problems, concepts, and methods—that is, the "rhetorical resources" that disciplines create and maintain in order to solidify their boundaries. Contests over the department's undergraduate curriculum have the potential to shape not only very practical matters like hiring priorities and new course creation, but also the distribution of *rhetorical resources*—namely, formulations of "English" as a discipline. One of the primary rhetorical resources in this case is control over the names assigned to different programmatic elements—concentrations, degrees, and so on—of the department.

As rhetorical attempts to construct a sense of collective identification, the names that an academic department chooses to apply to its programmatic structures stand in for larger arguments about the mission and the justification of the department. What Charles J. Stewart, Craig Allen Smith, and Robert E. Denton say about terms involved in social movement debates also applies to conflicts within academic departments: The terms we choose "play a role in determining sides of a conflict, specific views of reality, notions of right and wrong, and needed corrective action" (161). As points where social struggles occur as views of reality and notions of right and wrong are negotiated, the names we give to our pedagogical and scholarly endeavors provide important sites for examining how language intervenes between division and cooperation within academic units.

No doubt the most contested term we battled with in our proposed curricular revisions was the name "English." The disciplinary boundaries established through this term are tremendously volatile. While many scholars in different areas of English talk about "English Studies" in their scholarship—and by this phrase signify various textual specialties—the term "English," rather than the two-word name "English Studies," often remains the official name of academic departments. This official name, which omits the plural "studies" designation, reflects the pugnacity with which particular areas of study remain the expected focus of English departments. As many scholars have documented, the name "English" has, over the past century, come to equal "Literature," and an "English major" means a "Literature major," no matter how many times we refer to "English Studies" in our scholarship. While literary studies are by no means monolithic—the name "literary studies" in fact encompasses a wide array of texts and scholarly approaches to them—this area has been defined by some scholars within the specialty in a way that limits the scope of "legitimate" textual studies within English departments.

Such legitimating processes of definition are illustrated in some of the discussion that circulated within our department after we introduced our proposal. In response to the controversy, an ad hoc committee was formed to explore ways to revise the department's curriculum. As part of this exploration, the committee circulated a survey to faculty, asking for opinions about the missions and purposes of the English department. While several people envision a department devoted to language study broadly conceived to include literary studies, linguistics, composition and rhetoric, creative writing, and technical and professional writing, others expressed a belief that literary study is *the* business of English departments. In one survey response, a faculty member urged the adoption of a curriculum that would include the most possible literature courses. In another response, a faculty member recommended that the required "core" courses for the department should include only literary surveys.

Equations of "English" with "literary studies" result in part from the ways in which the term "English" is defined and structured by professional organizations that claim to represent practitioners in the field. Karen Fitts and Bill Lalicker point out that the MLA has the power to define and delimit "English" in a way that determines "what is central, what marginal; what's remarkable, [and] what's barely noticed" (428). Recent articles included in the MLA's *Professions* journal, Fitts and Lalicker argue, portray teaching writing as drudgery—as the work that must be endured before the teaching of serious and valuable "English" courses can take place. These understandings—or misunderstandings—of writing instruction make it very hard to create a larger, more inclusive understanding of the term "English."

A 2003 *ADE Bulletin* dedicated to discussing the English major similarly reinforces the centrality of a particular kind of non-utilitarian, aesthetic literary study in English departments. Addressing the doubts of many parents about the practical value of an English degree for their children, editor David Laurence suggests that "a specific and valuable sort of uselessness characterizes true engagement in the learning that serious consideration of literature uniquely affords"; yet, he continues, "part of the value of that specific uselessness lies in how useful it eventually shows itself to be in various walks of life. We are bound to be incurably ambivalent and conflicted on the subject of the 'practical value' of studying English" (5). Laurence makes an excellent point about the significant long-term practical value of studying literature: the descriptor "useful" too often is used to denote only immediately measurable and applicable skills, ignoring the practical benefits of long-range attitudes and habits of thought. Indeed, the practical value of what Peter Elbow calls "imaginative" language is often overlooked. As Elbow explains, "Imaginative language touches people most deeply; sometimes it's the only language use that gets through" (537). In other words, to achieve the effect we wish to have upon an audience—to accomplish a very practical goal—we need to be able to use imaginative language, and the study of literary texts is often a tremendously effective way to develop facility with such language. Yet Laurence's assertion about the "incurable ambivalence and conflict" surrounding the study of "English" rests on the assumption that an English degree does not include courses in rhetoric and composition or technical and professional writing, courses with more immediately identifiable practical value.

The name "English" is made to signify and exclude certain kinds of teaching and research in local contexts as well. For instance, the BA in English in our department had two concentration options at the time we proposed a new curriculum. One concentration was called a concentration in "English." This literary-intensive option was the "regular" concentration—the one that had no modifier and was identical in name with the department itself. The second concentration option was a Concentration in Writing. It is no coincidence that the concentration focusing on literature was called the concentration in "English" (and thus named the same as the department as a whole) while the Concentration in Writing was designated by a different term.

Underlying these attempts to identify what does and does not count as "English" are well-established assumptions about disciplinary unity—unity in purpose and mission. One of the clearest forms of what Gieryn calls "boundary-work" is the strategic act of defining a discipline's purpose or mission according to a principle of coherence that legitimizes one type of activity while delegitimizing others. In the case of "English," claims about the "core" of the discipline—and even claims that English is a unified discipline—create a cultural map of "English" that normalizes certain types of work while pushing other work to the margins. The components, and indeed the very concept, of a unified core for the English department, not surprisingly, became a site of heated discussion during our departmental wrangling.

As mentioned above, both concentration options started with a common "core" consisting of "Interpreting Literature," "Shakespeare," and three courses in literary history. Beyond this core, the two curricula diverged almost completely. Students in the "English" concentration took eighteen additional hours in literature plus one writing course, while the students in the Concentration in Writing took eighteen additional hours in writing (picking at their discretion from courses in creative writing, technical and professional writing, or rhetoric and composition) plus one non-writing course. In this Concentration in Writing, students did not receive systematic introduction to the concepts, questions, and methods of rhetoric or professional writing. One of our main goals in designing and proposing the RSPW curriculum, thus, was to ensure that students would have a "coherent experience" of these areas on some level. We proposed introductory courses in rhetoric and professional communication, plus a "capstone" seminar in which students would be asked to reflect on the whole of their major experience. In the disciplinary map defined by our proposal, we intended to provide a meaningful "core" experience for students interested in rhetoric and professional writing, a core that would replace the disjointed experience provided by the Concentration in Writing and a "core" of literature courses that made almost no reference to what those students would later encounter in their advanced writing courses.

Ironically, it was precisely our desire to create a "coherent" program that got us into trouble—not because other faculty rejected the idea of coherence, but because they disputed the principle of coherence that our curriculum proposed. In fact, nothing was more consistent than the argument by our critics that students should have a "shared experience" of some kind—but when these critics argued that students need a "coherent experience" of the English major, what they meant was the "coherent experience" that the curriculum, as configured before our proposal, provided: the fifteen hours of literature that existed in both concentrations. It was this core experience of literary study that our proposal threatened.

The idea that the department, rather than the concentration options available within the department, should be the level at which students share a common academic experience has been championed mostly, though not exclusively, by literature faculty, who have also been those most likely to refer to "English" as a *discipline* with various sub-disciplines (including creative writing, rhetoric, professional writing, linguistics, and so on). By contrast, many of the faculty in areas other than literary study have tended to refer to their own areas as "disciplines" existing within the English department, which itself serves as an institutional framework of related disciplines, each with its own core. Several responses to the faculty survey circulated by the ad-hoc committee suggest that English is not a discipline in the way other so-called disciplines, such as history and mathematics, are. Rather, English is an administrative structure that coordinates related but different disciplines. In this cultural map, "English" has an *institutional* reality, while rhetoric, professional writing, and so on have the more fundamental disciplinary realities. Working from this view of English as an institutional, rather than disciplinary, reality, some faculty have suggested that we investigate the possibility of establishing a College of English with different departments within that college.

The "disciplinary boundary-work" perspective illuminates how a group of generally fair-spirited, sincere, and intelligent people can hold and vigorously defend positions that appear irreparably at odds with one another. Departments of English are eclectic spaces, and it is not surprising that the residents of those spaces would depend upon different maps to help them make their way. It might be too much to expect to formulate a single boundary map that everyone can use. Absent that possibility, if we want to be able to refigure and expand maps that others have had a hand in drawing, it is handy at least to be able to read those maps effectively.

"ENGLISH" LESSONS

By enabling better understanding—better readings—of the different disciplinary maps operating within English departments, the boundary-work perspective can also enable more productive discussions within those departments. In this section, we highlight some communicative strategies that might lead to a process of change that is less fraught with territorial tensions. More specifically, we present some steps we might have taken—and that we think other faculty at other institutions may wish to take—to better prepare the way for proposed curricular changes. This is not to suggest that we could somehow have avoided all resistance or to imply that it's always in the best interest of professional writing and rhetoric faculty to remain housed within "English" departments (indeed, there are many successful, independent programs); rather, if we wish to maintain close departmental ties with our colleagues in literature—and, at this point, this is our goal at ECU—we have a responsibility to find out how best to communicate and justify our proposals to them.

Strategy #1: Discovering and Addressing Mistrust and Misunderstandings

One thing our story suggests for others who might be in a position to propose changes such as we did is the necessity to conduct—preferably before proposing curricular changes—the kind of meta-analysis of faculty alliances, faculty understandings of key terms, and faculty perceptions of disciplinary boundaries that we have conducted in this article. Discovering fault lines in faculty understanding of a department's identity and purpose is a critical first step toward productive change.

To be sure, we needed to better address our colleagues' mistrust of things "professional" and "technical." The attitudes of many assessment-focused bureaucrats toward the liberal arts has resulted in gut-level hostility on the part of some of our colleagues toward any program that proposes to teach communication that is in any way technical or business-related. Many of our colleagues resent the discourses of business and technical communication because these discourses are often used by those who want to "streamline" university budgets and to measure learning as quantifiable outcomes, despite the fact that, from a humanistic standpoint, much learning is not quantifiable. David Laurence's lament that the usefulness of literary study is often not readily apparent, at least not in the way that other kinds of workforce skills are, reflects this dissatisfaction with attempts to gather and report outcomes data about graduates of literary studies programs. Understandably, many of our colleagues in literature wonder, along with Richard Ohmann, "How can the complex things we most highly value be reduced to numbers?" (63). These colleagues—with justification—look skeptically on attempts to gather, analyze, and report data because such communication strategies have been used by assessment professionals to discredit and downsize academic programs in the liberal arts.

Some of the specific vocabularies of business, and thus of professional communication, have similarly fallen into disrepute among many of our literature colleagues. As Ohmann explains, "All in the arts and science . . . are likely to be put off by the ideas and language of business that have trailed along with accountability in its migration into the university" (63). Ohmann relates the details of a 1999 conference on "Market-Driven Higher Education," in which leaders discussed business management concepts such as " 'customization,' 'knowledge management,' 'just-in-time learning,' 'strategic partners,' [and] 'faculty management'" (63). Attempts to bring business management, and the predominant language of that management, into the administrative structure of the academy are threatening to faculty members who are not used to being "managed": "In short, when politicians or business people or trustees call for accountability in higher education, they are asking administrators to plan, oversee, and assess our labor," a process that academics, accustomed to or at least enamored of the idea of academic freedom, tend to resist. It is little wonder, then, that some of our colleagues looked with serious reservations on proposed curricular changes that gave a visible presence to the teaching of the discourses of business, even if our actual curricular aim was to teach the responsible, ethical use of such discourses.

While the "technical and professional" aspects of our proposal raised hackles, so too did its focus on rhetoric. As we discussed our proposed changes with our colleagues in literature and with some colleagues in other departments, we discovered that people often either do not know what rhetoric means, or they assume that they know what it means and that it is not good. On the one hand, some equate rhetoric with composition, and, since many faculty have been socialized within academic programs that see first-year writing as a stepping stone to bigger and better academic pursuits, the idea of giving rhetoric a prominent place in the undergraduate BA program seemed paradoxical. On the other hand are faculty who associate rhetoric with verbal trickery, with "empty" political talk, and with downright deception. Rhetoric, in these perceptions, does not merit serious scholarly attention. When the term "rhetoric," understood as verbal trickery, was combined with the term "professional" in the curricular structure that we proposed, some of our colleagues read the program as a training program in corporate deception.

Strategy #2: Publicizing What We Do

These perceptions of the fields involved in our proposed curricular revision reflect a reluctance to accept new areas of scholarship and teaching into the realm traditionally reserved for literary study. But, as we have come to realize, they also reflect the need for the architects of programs in professional writing and rhetoric within existing English departments to undertake a concerted campaign of educational publicity. The perceptions of some of our colleagues in literary studies about rhetoric and professional writing are inaccurate, but not necessarily because of territorial ill-will. Rather, these colleagues are reacting to their experiences of actual institutional conditions. So how do we change their views of what we do?

First, we need to try to bridge the conceptual gaps between the study of literature and the study of rhetoric and professional writing. To address these perceived gaps, we might employ some of the critical arguments put forth by well-known scholars of both literature and rhetoric. As Peter Elbow points out, scholars of both literature and rhetoric have argued for the commonalities of texts, regardless of where they might fall on a spectrum from "imaginative" to "technical" or "professional":

> Wayne Booth has made it clear that even literature has designs on readers—argues, does business.... [T]he tradition from Nietszche and I.A. Richards provides the opposite lens to help us nevertheless see that all language use is also an instance of poetics.... What's sad is that a discipline devoted to understanding language use should tend to restrict itself to one lens. (539)

Stressing such commonalities among texts might help alleviate the perception that rhetoric and professional writing are fundamentally different endeavors from the work of poets and novelists.

Secondly, we have come to realize that we need to illustrate the expanding theoretical frameworks within which professional and technical communication have developed—a development that many of our colleagues in literature are not aware of. Too often, our colleagues see courses in professional writing as handmaidens to other areas—business, engineering, science—courses that are components of curricula designed to make students more successful in other specialized fields. Yet, as we know, our courses have evolved to provide a much deeper education for our students. Once strictly service courses offered to majors from various parts of the university, writing courses in the technical professions, business, and the sciences have evolved into more or less coherent programs of study, exploring how certain kinds of specialized information can be communicated to those who need the information both within and outside of the technical or professional fields. At the same time, faculty in professional communication have worked toward carving out a niche for their research as well as their teaching. Although traditionally perceived as simply formulaic and practical, the research of educators and practitioners in professional writing has helped to define an endeavor rich in theory as well as practice.

Additionally, our colleagues need to realize the purposes and benefits—beyond marketable skills—of knowledge in rhetoric and professional writing. While some in the department will be persuaded of the importance of preparing students with "practical" writing skills, others simply will not be, not because they don't want our students to be employable graduates, but because they don't want them to *just* be employable graduates. This second group of colleagues mistakenly sees programs that focus on "professional" writing as primarily vocational rather than critical. Thus, we need to build into our publicity attempts examples of how instruction in professional writing, particularly when coupled with a rhetorical approach, goes beyond the mere transmission of technical or corporate skill. We need to reveal that our scholarship and pedagogy are not part of a callous endeavor to produce students with quantifiable workplace skills—rather, this instruction sensitizes students to the power of language, to the presence of propaganda, and to the ethical/humanistic concerns of communication in a variety of contexts, including the workplace.

At the same time, we need to explain how programs in professional writing and rhetoric can promote less-quantifiable cognitive goals—e.g. critical thinking skills—and, perhaps most importantly, can encourage the integration of these critical thinking skills into communication used in technical and professional settings. Isn't it better, we might argue, for our departments to teach students about the rhetorical impacts and the ethical consequences of writing in professional situations than to let them enter these endeavors without exposure to such considerations? Writing for business need not be part of an attempt to further the heartless desires of capitalism—in fact, education in professional writing might undermine these desires as students discuss the ethical, cultural, and social aspects of communication in business and industry. For English departments to cast off professional communication is for them to ignore the part they might play in encouraging students not to perpetuate oppressive corporate ideologies.

A good source to consult when considering how to explain the benefits and merit of a program in RSPW is Carolyn Miller's 1979 *College English* article, "A Humanistic Rationale for Technical Writing." Although this article is over thirty years old, Miller's arguments are still germane today. Teaching professional writing from a "flagrantly rhetorical approach," Miller argues, would, in fact, "present mechanical rules and skills against a broader understanding of why and how to adjust or violate these rules, of the social implications of the roles a writer casts for himself or herself and for the reader, and of the ethical repercussions of one's words" (617). "[A] course in scientific or technical writing" she continues, "can profitably be based upon this kind of self-examination and self-consciousness," thus furthering what Miller calls the "central impulse" of the humanities (617). Perhaps if we make these kinds of connections explicit for our colleagues, we can alleviate their fears.

Strategy #3: Reviving the "Practical"

While it is essential to alter misunderstandings of professional writing that see it merely as a vocational endeavor, we might also benefit from attempts to resuscitate the practical within English departments. Ellen Cushman has suggested that "English studies *must* avoid simple vocational training: the uncritical, unexamined acquisition of skills that apply mechanically to workplace production and distribution of information, products, and services." But, she continues, this does not mean that instruction in modes of writing that relate to these activities should be abandoned or held in lower esteem than the study of literary textuality. The key, she explains, is to separate the teaching of writing for the purposes of vocationalism (writing for career advancement) and the teaching of writing for the purposes of utilitarianism (writing to get things done). "[V]ocationalism," she clarifies, "should be differentiated from utilitarianism. . . . Utilitarian knowledge can be made and put to use by well-rounded, knowledgeable, socially conscientious students, citizens, and professors who together try to better the public and private institutions they are both critical of and reliant on" (213).

Scholars in English studies need to be made aware that there is important middle ground between selling out to corporate America and providing critical instruction in efficacious knowledge. Our students will find it extremely difficult to survive economically if we fail to prepare them to communicate in contemporary workplaces and other public settings. As Cushman puts it, "Any reform of English studies must consider how ultimately the knowledge made in English can be of economic and social value, can accrue cultural capital, and can help its bearers accrue symbolic capital" (213). For graduates with degrees in English to implement—in other words, to gain the symbolic and economic capital to put into practice—the kinds of social changes we might wish to see, those graduates will need instruction in utilitarian kinds of knowledge. They will need the rhetorical skills to communicate effectively in professional contexts. Teaching in this kind of utilitarian framework might provide "skills" for communicating in professional contexts, but it would do so with an undercurrent of critique—the kind of critique that literary writers have long promoted through poems, novels, and other forms of literature.

Of course, we will not be able to change everyone's view of the proper sphere of the English department. Those who hold that certain kinds of writing are inherently superior to others or who come to the table with other departmental agendas will not necessarily care what we have to say. But, in any attempt to elevate the presence of rhetoric and professional writing within English departments, those colleagues should *not* be our primary intended audience. Instead, we need to focus our persuasive efforts on colleagues who are *legitimately* skeptical of our proposals, rather than immovably against them. Even with such focused approaches, we might find that the negative views of rhetoric and technical and professional communication are too strong. Indeed, where space, finances, demand, and political climate permit, several scholars and teachers of professional writing and rhetoric have found the best solution to departmental conflicts to be the establishment of separate academic departments. If we wish to remain—for whatever reasons—within a broad-scope department like ours at ECU, however, we need to promote what we do in our theory and in our practice.

Strategy #4: Rethinking the "Core"

In addition to promoting what we do, curricular proposals such as ours call for advanced discussion about the purposes and the location of "core" courses. Such discussion had taken place in the past in our department, but those discussions were several years removed. It may have made the process of changing the undergraduate curriculum less antagonistic if we had engaged in these conversations shortly *before* putting forward our proposal. How one understands the purposes of a "core" will of course impact one's response to curricular proposals that configure a core in a particular way. In their survey responses to the aforementioned ad hoc committee, faculty proffered three major understandings of what a "core" in the English department should provide for students: 1) a common set of skills needed for academic, professional, and/or civic achievement; 2) a common body of knowledge, understanding of which should characterize English majors; and 3) a combination of skills and knowledge that together will prepare students for academic, professional, and/or civic achievement. Obviously, each of these understandings would lead to a significantly different "core" of required courses. Exploring the purposes of a "core" and imagining different options available for locations of "core" knowledge within a degree program might have made it easier for us to present our plan to colleagues without seeming to threaten what they value.

At the time the "core" is discussed, it would also make sense to provide alternative visions of "core" knowledge in English departments. Such alternatives can be garnered from other departments and from a variety of scholarly publications. Jonathan Culler, for instance, provides some ideas for how to reimagine the concept of the "core" in an English department in such a way that students do have some common experience across the different areas of English Studies but that does not privilege one subject area over others. Although Culler begins his piece "Imagining the Coherence of the English Major" with a three-and-a-half-page discussion of how to create a unified English Major as a literary degree, he goes on to acknowledge that this kind of literary-based coherence comes at a significant cost:

> The major drawback may be, however, that this approach defines the English major as a literature major, neglecting all the other things that English

departments have come to do—including the study of other sorts of writing; the practice of writing itself, whether expository or "creative," as we oddly call it; and the study of other cultural practices, such as film and television. (9)

The concluding page and a half of the article articulate a vision of an English department that would include literature and all these other things. Culler's proposal focuses on developing the "unity" of the English major through different abilities and habits of mind that various courses might cultivate in students. More specifically, he proposes that

> English departments attempt to define the sorts of learning that we think ought to take place and that might be achieved in the English major. For instance, an English major might include literary and rhetorical analysis, historical analysis, social analysis, cultural analysis, cognitive and moral analysis, and the practice of writing. Here, I think, we have distinct sorts of analytic practices that students can acquire, all in the broad structure of the English major; the coherence of the major would lie in its attempt to provide instruction in this full range of practices. (10)

This arrangement would ensure that students in the sprawling English department have common abilities, even if they do not all graduate with the same content knowledge.

Strategy #5: Highlighting Institutional Realities

We might also explore the strategic, yet admittedly materialistic, "power in numbers" argument at the same time that we suggest ways in which our studies and goals overlap with those of our literature colleagues. Pat Sullivan and Jim Porter have mapped out the spaces occupied by professional writing in English departments and explored the struggles faced by this relatively new terrain of professional writing, arguing that the "development of professional writing as an academic entity signals a key conceptual shift: from the traditional notion of writing as ancillary to some other subject matter . . . to a recognition of writing as a discipline in its own right . . . " (405-06). They conclude that professional writing may be at home in "English," but question whether English departments can afford the resources to support these programs. Perhaps more significantly, they also ask if English departments can afford not to support these programs. More recently, David Downing has argued that administrators, under pressure to reduce expenses, "are the only ones to gain from internecine warfare among competing subdivisions. In the end, isolation makes any small unit or program more vulnerable to administrative surveillance" (31). Such an argument, however, needs to be made strategically. Because of associations between professional writing and business interests, some colleagues may resist this argument, seeing it not as a practical reality but as yet another way that they are being pressured to submit to the interests of those driven by assessment and efficiency.

Strategy #6: Discussing Names

The resistance we met with also suggests that a prior or concurrent discussion should have addressed the official name currently assigned to our department. Our proposal might have fared better if we had considered a new departmental name—a more appropriate combination of terms that reflects the scope of the work that actually goes on within "English" departments. Perhaps the term "English" is too laden with previous meanings and assumptions to be useful as a signifier with which the commonalities of the current-day English department can be represented. As one respondent to the departmental survey put it, "I don't think a student should get a degree in English without having a substantial background in literature. Call the degree something else if necessary, but don't call it a degree in English." How about the Department of English Language and Literature? Or, to give writing an even more visible presence, the Department of English Literature, Language, and Writing Studies? These names identify and thus privilege multiple strands of research and pedagogy, better reflecting faculty expertise and, perhaps more importantly in terms of attracting students, explaining more clearly for undergraduates what they can study and learn within the department.

While reconsidering a bifurcated departmental name, it would also perhaps be worthwhile to ask why there needs to be one, unified mission and only one word (to reflect this supposed unity) in our name. Disciplines, Michel Foucault suggests, are not unified bodies of knowledge but disparate ones. This view of disciplinarity, Craig Dionne and David Shumway suggest, "conflicts radically with our expectations, and it should lead us to wonder where the criterion of unity comes from and why it should be applied" (6). The ability of the department to function together and make the best use of the various talents of its teachers, researchers, and students is perhaps best served by acknowledging that we don't all do the exact same thing and that we don't all hold the same goals to be equally important. As Elbow suggests, perhaps "a discipline can be even richer and healthier if it lacks a single-vision center. A discipline based on this multiplex model can better avoid either-or thinking and better foster a spirit of productive catholic pluralism" (544). Accepting a multivalent construction of "English" would also be an acknowledgement of the reality that the discipline has never, in fact, been fully congealed around one methodology or body of knowledge. No English department—since there have been such departments—has ever been smoothly, wholly unified. If there was such unification, it would most likely indicate stagnation.

The process of negotiating what "English" means and the lack of understanding exhibited by some of our colleagues—and here we stress the word "some" because there has been significant support for change in the curriculum—have not always been pleasant. But we are happy that the conversation is underway. It's a necessary process that many other departments have undergone and that still others have yet to begin. Too, there have been moments of productive cooperation. Comments from many faculty show a desire to structure the department as one that welcomes an expansive array of approaches to texts and a multi-faceted understanding of the kinds of writing that might fit within a diverse department. One survey respondent, a literature specialist by training, reminded readers of what our departmental mission statement says. This mission statement, despite its moments of elitism, presents the department as an open space that, as a matter of course, values language, writing, *and* literature, all of which are "integral" to the department:

Members in the department share these assumptions:
1. Language is fundamental to human nature and is at the heart of intellectual life.
2. Literature permits us to engage our consciousness with singular keenness, profundity, and pleasure.
3. Writing engenders social, cultural, economic, and political vitality.
4. Language, literature, and writing are integral.

This statement reflects Robert Scholes's revised model of English studies, putting "textuality" in all its forms, rather than only literary works, at the center of our endeavors. Perhaps, as our colleague suggests, we might revisit our mission statement and rededicate ourselves, as a department, to textuality.

EPILOGUE: THE STAGE FOR FUTURE CHANGE

Downing, Hurlbert, and Mathieu suggest that "when taking collective action, moments that feel like failure may have future effects we cannot know or imagine. For example, if a group plans an ambitious new curriculum and it fails to be implemented, that process might have succeeded in other ways: bringing people together to highlight tacit departmental divisions . . . or setting the stage for future change" (13). Since we began working on this account of the turmoil that ensued when we attempted to alter the curricular structure of the department, we have witnessed the kind of "future change" that can eventually result from a moment—in this case the demise of our proposed RSPW concentration—that feels like failure. As we mentioned earlier, in response to the issues raised by our proposal, an ad hoc committee was formed to conduct a survey of faculty views on the nature of an English degree. The committee was further charged with developing a revised undergraduate core curriculum based upon these views. The work of the ad hoc committee has resulted in several significant changes to the department's undergraduate degree, changes that were unanimously approved by departmental faculty. The two most significant changes, which went into effect in spring 2005, are as follows:

1. The core group of courses required of all majors in the department was revised to give some presence to classes dealing with rhetoric, professional writing, and English language study. At the same time, the number and specificity of core requirements was reduced, thus allowing students more opportunity to explore the variety of specialties within our diverse department. The new core is elaborated in the table below.

Old Core (required of all majors in the department before Spring 2005)	Revised Core (required of all majors in the department beginning Spring 2005)
• ENGL 2000. Interpreting Literature • ENGL 3000. Lit in English to 1700 • ENGL 3010. Lit in English, 1700-1880 • ENGL 3020. Lit in English, 1880-Present • One Shakespeare course (Tragedies, Comedies, or Histories)	• One Historical Survey (selected from a variety of offerings in Literature pre-1700) • One Historical Survey (selected from a variety of offerings in Literature post-1700) • One Shakespeare course (Tragedies, Comedies, or Histories) • One Language Study Course (chosen from a variety of courses in Creative Writing, Linguistics, Rhetoric & Composition, and Technical & Professional Communication)

2. Separate, named concentrations within the BA in English were eliminated. Instead of a BA in English with options for Concentrations in Writing or Literature, the department now simply offers a BA in English for all students. As they pursue this degree, students are expected to work with their faculty advisors to create a curriculum of upper-division courses that will best meet their interests and advance their future plans. While the removal of concentrations did not give RSPW an official, named presence in the department (something we'd initially hoped our proposal would accomplish), the change provides opportunities for students to expand their studies in RSPW in ways that were not possible under the previous structure of concentrations.

At the same time that the core was being revised, several new courses in rhetoric and professional writing were added to the department's regular offerings, including two courses—Introduction to Rhetorical Studies and Introduction to Professional Writing— intended to introduce students to RSPW as an integral part of the English department. While we did not find success with our initial proposal, we are encouraged by the more visible presence we now have in the curriculum and by the attendant possibilities for collaboration among the various specialties within our large and diverse faculty.

WORKS CITED

Culler, Jonathan. "Imagining the Coherence of the English Major." *ADE Bulletin* 133 (Winter 2003): 6-10.

Cushman, Ellen. "Service Learning as the New English Studies." *Beyond English Inc: Curricular Reform in a Global Economy*. Ed. David B. Downing, Claude Mark Hurlbert, and Paula Mathieu. Portsmouth, NH: Boynton/Cook, 2002. 204-18.

Downing, David B. "Beyond Disciplinary English: Integrating Reading and Writing by Reforming Academic Labor." *Beyond English Inc: Curricular Reform in a Global Economy*. Ed. David B. Downing, Claude Mark Hurlbert, and Paula Mathieu. Portsmouth, NH: Boynton/Cook, 2002. 23-38.

Downing, David B., Claude Mark Hurlbert, and Paula Mathieu. "English Incorporated: An Introduction." *Beyond English Inc: Curricular Reform in a Global Economy*. Ed. David B. Downing, Claude Mark Hurlbert, and Paula Mathieu. Portsmouth, NH: Boynton/Cook, 2002. 1-21.

Elbow, Peter. "OPINION: The Cultures of Literature and Composition: What Could Each Learn from the Other." *College English* 64.3 (May 2002): 533-46.

Fitts, Karen, and William B. Lalicker. "Invisible Hands: A Manifesto to Resolve Institutional and Curricular Hierarchy in English Studies." *College English* 66.4 (March 2004): 427-51.

Gieryn, Thomas F. *Cultural Boundaries of Science: Credibility on the Line*. Chicago: University of Chicago Press, 2000.

Laurence, David. "Notes on the English Major." *ADE Bulletin* 133 (Winter 2003): 3-5.

Miller, Carolyn R. "A Humanistic Rationale for Technical Writing." *College English* 40.6 (February 1979): 610-17.

Ohmann, Richard. "Accountability and the Conditions for Curricular Change." *Beyond English Inc: Curricular Reform in a Global Economy*. Ed. David B. Downing, Claude Mark Hurlbert, and Paula Mathieu. Portsmouth, NH: Boynton/Cook, 2002. 62-73.

Russell, David R. "Institutionalizing English: Rhetoric on the Boundaries." *Disciplining English: Alternative Histories, Critical Perspectives*. Ed. David R. Shumway and Craig Dionne. Albany, NY: SUNY P, 2002. 39-58.

Scholes, Robert. *The Rise and Fall of English Studies: Reconstructing English as a Discipline*. New Haven: Yale University Press, 1998.

Shumway, David R., and Craig Dionne. Introduction. *Disciplining English: Alternative Histories, Critical Perspectives*. Ed. David R. Shumway and Craig Dionne. Albany, NY: SUNY P, 2002. 1-18.

Sullivan, Patricia, and James E. Porter. "Remapping Curricular Geography: Professional Writing in/and English." *Journal of Business and Technical Communication* 7.4 (1993): 389-422.

REVISING

5 Smart Growth of Professional Writing Programs: Controlling Sprawl in Departmental Landscapes

Diana Ashe
Colleen A. Reilly

INTRODUCTION

At our first departmental planning retreat in January 2004, we gave a twenty-minute presentation in front of our assembled colleagues. Our goal was the creation of an administrative position for a professional writing coordinator, a course release for that position, and some other additional resources. To that end, we detailed the constraints under which our Professional Writing program had been operating: we had 108 students—fully one-third of the department's majors—enrolled in our program, yet we were the only two tenure-track faculty in an English department of twenty-five fully dedicated to our PTW major and certificate program at UNC Wilmington[1]. In addition, we detailed some of the administrative duties involved in meeting the needs of the program in its first four years, including establishing and obtaining university approval for the program; communicating with prospective, current, and former students; auditing graduation requirements and distributing certificates; scheduling all PTW courses every semester; recruiting, screening, interviewing, and mentoring part-time faculty; applying for and carrying out grants for curriculum development within the program; chairing the professional writing committee; and advising prospective and current students in the program. We felt that we could no longer carry out these ever-increasing duties without harming our research agendas (and, thus, our tenure hopes), so this presentation was of critical importance.

Once we finished stating our case, we opened the floor for questions, and the hand of a colleague with primary interests in literature was raised. She asked quite earnestly, "What exactly *is* professional writing, anyway?" We responded by giving a few working definitions and moved on to more specific concerns about the changes we were proposing, but the question had tremendous impact on us. Part of the impact derived from the difficulty in answering this question briefly. More significantly, though, being confronted by this question impressed upon us both the lack of understanding among some of our colleagues for what we do and the position of our program as something of a foreign entity within the English department, where, to our knowledge, all other subdisciplines were accepted on their face as comprehensible and appropriate. The question speaks to scholarship in professional and technical writing that raises the issue of how and even whether to define professional and technical writing. Pithy definitions have been developed, such as David Dobrin's assertion that "Technical writing is writing that accommodates technology to the user" (242), which is still referenced positively in introductions to the field and discipline (see Lutz & Storms). However, other scholars, such as Jo Allen, object to restrictive definitions like Dobrin's, especially as they are not based upon systematic study, and are often used to exclude certain types of work or other writers from the field. Allen cautions against creating definitions and argues that it is better for us to "keep our field intact—with our impressionistic, experience-based ideas of what technical writing encompasses—than to succumb to simplistic or exclusionary definitions that separate us from one another" (77). Recently, many seem to agree with scholars such as Spilka, who argues that the diversity of definitions of professional and technical writing indicates that the field is healthy, characterized by "diversity, fluidity, a contextual nature, interdisciplinarity, and multiplicity in terms of career paths and specializations" (102-3).

In our local departmental environment, we have experienced both a continual request for definitions of our field as well as the objections that follow when our descriptive definitions of our major and the field contain something objectionable, such as references to technologies or workplace contexts, or exclude something that colleagues outside the field perceive as belonging to it, such as journalism. In January 2006, we were again asked to define professional and technical writing at a series of meetings that led up to another departmental retreat in February 2006. Our discussions with colleagues at these meetings and informally in the halls reemphasized that operating a professional and technical writing program within an English department entails more than all of the duties we list above; it also entails operating within a collegial and organizational context, one that reaches out to and is reached by stakeholders and community members at every possible turn. Unless our colleagues share our understanding of our curriculum, mission, and goals—something we continue to struggle with—we will never achieve our hopes for the program and our students will always be underserved. Conversely, our goal of offering a consistent and carefully balanced set of courses with the strongest faculty and most current resources we could muster will not be truly successful until we acknowledge and understand our program's unique departmental and university environments.

In the pages that follow, we borrow concepts from systems thinking and from the smart growth movement to conceptualize the necessity and potential for situating professional and technical writing programs interdependently within their larger organizational contexts. Systems thinking offers clear guidelines for designing and maintaining programmatic operations, guidelines that delineate specific goals and actions on the way to creating sustainable and resilient programs. Smart growth planning demonstrates the specific actions within each of the systems thinking guidelines that strengthen and clarify the relationships between programs, stakeholders, and communities. Through our discussion, we show how we have tried to plan the future of our program with the help of these two matrices, how these principles have influenced our attempts to define, flexibly and from a systems perspective, professional and technical writing as a field and as a program within our English department at our university, and how we have tried to use these principles to articulate and demonstrate more clearly the connections between our work and that of our colleagues as well as better encourage them to participate in and understand the work that we do. Our successes as well as our failures provide insights into the usefulness of systems thinking and smart growth as bases for directing programmatic growth and expansion in professional writing.

APPROACHES TO PROGRAMMATIC GROWTH IN PROFESSIONAL WRITING

Writing programs in general and professional writing programs in particular often fit uneasily within humanities departments, such as English, despite often originating in those departments and sharing many characteristics and goals with other humanities disciplines (Di Renzo, Rutter). Professional writing's epistemological and methodological connections to English can be seen as tenuous (Hocks, Lopez, and Grabill), and professional writing programs may have more success obtaining resources and finding support for collaborative scholarship, for example, in departments of business or technology (Davis). Institutional circumstances, however, may make it impractical for programs to relocate in other areas of the university and, therefore, growth and development strategies must be developed that work within institutional limitations and realities.

Numerous scholars and writing program administrators in professional writing advocate interdisciplinary approaches to program growth and development in order to prepare students broadly to work with information and communication technologies and gain expertise in subject areas outside of the humanities. In addition to competency in writing and rhetoric, scholars note that professional writing students benefit from training in computer sciences, graphic design, and organizational communication, which they can best and perhaps only get from other departments, depending on the expertise and size of program faculty (Blythe). In certain institutional contexts, creating professional writing programs as extra-departmental, interdisciplinary structures provides the best means for providing educational options for professional writing students; supplementing the skills of the professional writing faculty, who often number only a few; and gaining access for students to more technological resources than small programs in English departments may be able to provide (Blakeslee; Blythe; Andrews & Worley). Forging partnerships with and gaining participation from faculty in other departments can integrate professional writing into the broader university community, which can in turn provide exposure and stability to the program. Such integration can be accomplished using the model of WAC/WID programs already in existence and, in fact, a number of scholars advocate allying professional writing with WAC programs (Bosley; Hocks, Lopez, and Grabill). Other approaches to program development emphasize the work that needs to be done at administrative levels, including compact planning, which focuses on setting specific, incremental goals for the program and gaining administrative support for those goals (Allen).

A number of the approaches to program growth and development that we surveyed highlight the importance of and problems with creating a space within traditional academic structures, like English departments, for technical and professional writing programs (Hocks, Lopez, and Grabill), whose interdisciplinarity and focus on workplaces and technologies are not always easily accommodated by traditional notions of discrete departments and the concerns of humanities disciplines. The focus on space is by no means accidental, for academic units, including departments, schools, and universities as a whole, are organic entities sharing attributes of biological and environmental systems. Addressing the space and environmental issues raised by many developers of professional and technical writing (PTW) programs requires systems thinking, as we explain in the next section.

SYSTEMS THINKING AND ACADEMIC ECOSYSTEMS

Imagining our universities, departments, programs, students, and faculty as part of an academic ecosystem has both utility and precedent. Systems thinking dominates in contemporary scientific endeavors, putting emphasis on the interdependence of relationships between organisms and their environments. Our work as PTW program administrators is no different; resource allocation and day-to-day challenges may dominate our thinking, but relationships constrain or support our success.

Sydney I. Dobrin and Christian R. Weisser define ecosystems as "groups of organisms which function together in a particular environment (physical and chemical) and exchange energy within the system in order to metabolize, grow, and reproduce" (73). Dobrin and Weisser have put together volumes on the link between ecosystems and writing systems, and the connection has been touted for more than twenty years. Marilyn M. Cooper, who made the most widely-cited early suggestion of the potential of an ecological approach to composition, still emphasizes the idea that "the systems that constitute writing and writers are not just like ecological systems but are precisely ecological systems, and that there are no boundaries between writing and the other interlocked, cycling systems of our world" (xiv).

Extending the link between ecosystems and writing systems, we suggest that academic departments are ecosystems of their own, and that by thinking of them in this way we can highlight the spatial, geographic, and relationship aspects of academic units and the importance of considering these elements for the growth of programs within these units. Michael Weiler and W. Barnett Pearce use the term "rhetorical ecology" to describe viewing public discourse as "a kind of ecosystem in which various individual discursive subsystems interact in relations of conflict and mutual dependence" (14). Likewise, in the academic department, special interests must interact over curriculum, instruction and departmental resources.

Our role in this ecosystem is constantly changing and tends to provoke reactive changes in the roles of other members. As Weiler and Pearce suggest,

> Rhetors are forced to act within the confines of the ecosystem, and their discourses must reflect the web of relationships among its species and its surroundings. But as the rhetorical ecosystem evolves, as any living thing must, so too do its discursive possibilities, and within the system there is ample room for authorial creativity and cleverness (15).

The space within the department or university ecosystem for authorial creativity and cleverness offers program administrators opportunities for building programs that have internalized certain survival skills. Survival skills in our case would include careful planning for the inevitable changes that occur in our rhetorical ecosystems. Because "[c]ontext both fits rhetorical action and is reconstructed by it" (15), our decisions as administrators change the system and all of the relationships it affects and is affected by.

DEVELOPING ACADEMIC ECOSYSTEMS THROUGH SMART GROWTH

A number of fields in the humanities, including ecocomposition, rhetorical ecology, ecocriticism, and ecofeminism, look to ecosystems as foundational constructs. Outside the humanities, urban planners, political scientists, and sociologists use environmental science in other ways that can inform our thinking and bring a systems approach to program planning, development, and adaptation. A recent and particularly useful systems-based approach, smart growth, involves the application of broad-based systems thinking to land-use decisions and community development. Smart growth offers concrete strategies for handling growth and change that can be used productively to address the concerns of new Professional and Technical Writing programs. These strategies combine strategic planning, environmental awareness, and political negotiation as opportunities for identity construction and chances to demonstrate the appropriate place of the PTW program within the department and the university. In the face of fears that PTW is the academic equivalent of urban sprawl, the language and strategies of smart growth can help us to develop and strengthen our programs in ways that are in the best interests of the department, the university, and the community.

On the surface, smart growth principles may seem distant from the needs of PTW programs because they refer specifically to physical spaces and environmental concerns. According to both major smart growth coalitions in the United States, the Sustainable Growth Network and Smart Growth America, the main principles of smart growth are as follows:[2]

1. Create a range of housing opportunities and choices
2. Create walkable neighborhoods
3. Encourage community and stakeholder collaboration
4. Foster distinctive, attractive communities with a strong sense of place
5. Make development decisions predictable, fair, and cost-effective
6. Mix land uses
7. Preserve open space, farmland, natural beauty and critical environmental areas
8. Provide a variety of transportation choices
9. Strengthen and direct development towards existing communities
10. Take advantage of compact building design
    (Sustainable Communities Network, "About Smart Growth")

When considered through the lens of the academic program, however, the basic goals of this kind of systems thinking can serve as a framework for program administrators when setting goals for program design, program development, outcomes assessment, and strategies for anticipating and managing change.

According to Joseph Fiksel, resilient systems feature diversity, efficiency, adaptability, and cohesion. In order to design a resilient system, Fiksel recommends "identifying system function and boundaries, establishing requirements, selecting appropriate technologies, developing a system design, evaluating anticipated performance, and devising a practical means for system development" (5330). As Fiksel indicates, resilience leads to and fosters sustainability, which "is not an end state that we can reach; rather, it is a *characteristic* of a dynamic, evolving system." In order to foster their own resilience and that of the academic ecosystems of which they are a part, administrators of PTW programs can enact Fiksel's recommendations for sustainable development, which, as Fiksel argues, can be employed at any point in a system's development to alter its course (5334). In the remainder of this chapter, we will demonstrate how Fiksel's critical actions for designing a resilient system and sustainable growth, combined with the ten principles of smart growth from the Sustainable Communities Network offers strategies for building successful PTW programs that flourish within the departmental and university ecosystems where they have taken root.

IDENTIFYING SYSTEM FUNCTION AND BOUNDARIES

As a PTW program begins to grow, it is inevitable and essential to hold discussions about the direction and amount of sustainable growth possible in light of current resources and other institutional constraints. In order to lead and initiate such discussions, we felt that it would be essential to have one designated program coordinator who could be the official spokesperson of the program in discussions and negotiations with other faculty and administrators. One of our early successes was to get departmental approval and chair support for the position of coordinator of professional writing, which came with a list of duties, including permanent membership on the departmental steering committee, as well as a course release. Until we earned tenure, we shared the position, alternating years, in order to allow both of us to gain formal administrative experience central to developing sustainable programmatic growth.

In some cases, successful undergraduate programs, in particular, face departmental and administrative pressure to grow beyond capacity in order to attract more majors, serve growing numbers of interested students, and even create bridges between the community and private industry, something that administrators may uncritically see as the role of PTW programs. In order to handle growth, negotiate external pressures, and design a program that serves the entire university, reflection about the function of the program, including its goals and strengths, is essential and can help program administrators argue successfully for holding expansion and development to manageable levels.

In setting boundaries for a program to avoid taking on too much too soon, program administrators may draw upon smart growth principles six and ten, which advise: "Take advantage of compact building design" and "Mix land uses." The former recommends that development be compact, make efficient use of space and resources, build upon a strong foundation, and keep growth under control in order to avoid overtaxing resources. Likewise, the latter principle of "Mix land uses" highlights the importance of integrating the use of resources by commingling different populations in the same environment and maximizing the populations who benefit from the available services and resources that they need.

In practical terms, by thinking of PTW programs as part of the departmental and university landscape, program administrators can "Take advantage of compact building design" by directing growth in service of the stated goals of the program and of the department as a whole. For programs with few faculty that, like ours, are housed within an English department, the programmatic and departmental goals should be integrated and reconciled so that the PTW program can take advantage of the department's course offerings. Courses in rhetorical theory, essay writing, or journalism, for example, might be outside the strictest province of professional and technical writing, but can help students to build strong writing and rhetorical skills and supplement the more specialized knowledge in PTW that only a few faculty can provide. By taking fullest advantage of existing departmental course offerings, a sufficient number of courses can be offered within human resource constraints. Overall we have been quite successful in actively recruiting faculty outside of professional writing to teach such courses and to develop special topics courses related to writing and rhetoric, some of which, like Writing about Film, have become regular courses that appeal to our students as well as students in our university's burgeoning film studies program. However, we must acknowledge that a very small number of faculty who are skeptical about how our program fits within the humanities mission of the English Department refuse to teach courses that are within their areas of expertise because they are listed as professional writing courses. Nonetheless, over the past two to three years, we believe that we have effectively enlisted existing departmental personnel to offer a wide range of courses for our students and, simultaneously, to gain more support and understanding from other departmental faculty for our programmatic mission.

Programs can also efficiently use resources and shore up their foundations by avoiding taking on too much and focusing their programs around the specialties and backgrounds of existing personnel rather than trying to accomplish theoretical goals based on ideals gleaned from scholarship or past experience. For example, our program incorporates some journalism-oriented courses into the base requirements for our major. We have a number of talented part-time faculty to teach those courses who are trained and experienced journalists and no other department on campus teaches print journalism, so there is a significant student demand. Furthermore, the university has recently created an interdisciplinary minor in journalism that we can support and participate in. We include journalism in the professional writing curriculum despite the fact that none of the tenure-track faculty in PTW were trained in programs with such a focus, and it violates our instincts and strict understandings of our field to some extent. However, to attempt to build a program around only specialized courses, like Writing for the Computer Industry, would be currently spreading ourselves too thin, weakening our foundation and increasing our horizontal rather than vertical growth, making us less able to offer our students the ability to pursue some subjects in depth through a range of upper-level courses. Additionally we would be resisting the interests of our students, many of whom are interested in studying journalism and working as writers for local and national publications.

Keeping growth compact and focused further helps in the efficient use of resources, both human and technological. Concentrating on specific goals can make best use of both tenure track and part-time faculty by allowing them to concentrate on teaching classes in their specialties, which prevents them from becoming fragmented and overworked by continually having to learn new subject matter. Additionally, limited resources for acquiring technologies such as software can be spent in targeted ways rather than be used to acquire the latest tools in areas far afield of the central goals of the program. Finally, the goal of compactness may extend to considerations over the types of degrees and other credentials the program can award. Successful undergraduate programs such as ours, which attract many majors, may be pressured to expand to the M.A. level or offer courses to private industry before they are ready for this sort of expansion. Until additional qualified faculty and other resources make meeting the needs of undergraduate majors less of a struggle, referring to the benefits of compactness and a solid foundation highlighted by this smart growth principle can help program administrators to articulate resistance to premature growth.

Within a focused and targeted PTW program, administrators can be guided by the smart growth principle which advises "Mix land uses" by making the program appealing to a diverse population of students. This can be accomplished by providing ways for students with different majors to incorporate aspects of PTW into their programs of study. For example, English majors can obtain their degree in the forty-two-semester-hour major in the Professional Writing track, while students majoring in science or business can pursue a twenty-one-semester-hour Certificate in Professional Writing. Both the track in the major and the certificate require only a limited number of specific courses (two named courses and two from small groupings of courses for the major and three named courses and one from a small group for the certificate), allowing students with specific interests to select courses that relate to their own academic and career goals. This flexibility appeals to many students in and outside of English studies who are interested in writing and wish to add a formalized writing credential to their academic profiles and this appeal is supported in part by our growth in majors, up from 50-60 at the start of the program to 112 as of December 2005. Because we attract students from a variety of backgrounds in our courses, including our Introduction to Technical Writing, Writing about Science, and Writing and Technology, we provide an interdisciplinary learning experience including a diversity of perspectives and talents in the classroom.

In order to make the program accessible and palatable to a range of students, including non-traditional students, choices about the scope and direction of the program and courses have to be made with a number of often conflicting audiences in mind. For example, we require an internship for our certificate program but not for the major, as the internship may pose an obstacle for some students who want to pursue the PTW major. In addition to the minimum grade average required for an internship (which helps to ensure that only our strongest students are representing the university in this way), some non-traditional students, students with children, and students who already work full-time jobs may have logistical difficulties completing this requirement, so we incorporated it into only the optional certificate. Many of our PTW majors obtain the certificate as well, but occasionally we have students who are unable to do so due to work or family pressures. Additionally, many students in Communication Studies also pursue our Professional Writing Certificate or even double major in Professional Writing to hone their writing skills and help them further their career goals, particularly in print journalism.

ESTABLISHING REQUIREMENTS

In order to prepare the PTW program for growth, it is important to understand the program's current status and what would be required for growth, where opportunities exist for development, and what sorts of additions would most benefit students. Prior to planning for growth, including adding new courses and hiring new permanent faculty, program administrators need to understand and analyze the present level of resources and direct growth in sustainable directions.

The smart growth principle seven, "Preserve open space, farmland, natural beauty, and critical environmental areas," is relevant in determining what is present and what will be required in order to grow the program in productive and useful ways that capitalize on the strengths and address the gaps in the current curriculum and structure. Based on this principle, growth should preserve open space or flexibility, be redirected to existing communities, and help in removing development pressure. In planning new hires, should lines become available, candidates should be sought who both complement what is currently working and provide additional expertise related to targeted growth areas. For example, we recently hired a fourth specialist in professional writing who specializes in science and medical writing, which can help us to serve and speak to the large number of biology and marine biology students and growing numbers of nursing students, by state mandate, at our institution. Furthermore, although the fit is outside of strict definitions of PTW, we have agreed to assist the department in pursuing a future hire in journalism to serve the large numbers of students interested in that area in our program as well as our institution's new interdisciplinary minor. Preserving what is unique about our institution's offerings is, in this instance, more critical than delineating a textbook PTW program.

Preservation also extends from the program to the departmental level. The PTW program should be flexible enough to help to bolster what is good and useful in the department outside of the program and in related departments so as to integrate PTW and other areas. For example, our PTW program requires a significant number of courses from literature (nine to twelve semester hours) and allows students to take related courses in the creative writing department and count them toward the program. Such crossover preserves what is useful in the established programs in English—while simultaneously making it possible for our literature colleagues to continue to teach upper-level classes in their fields despite growing enrollment in PTW—and creative writing by drawing students to those courses while decreasing our development pressure, providing our students with more options and depth without overtaxing our permanent and part-time faculty in PTW. While some may question the relationship between other fields, such as creative writing, and PTW, at our specific institution the creative writing program is extremely popular and nationally renowned and, thus, working with them benefits us both. Another way to preserve open space is to build in enough elective credits that allow the program to easily adapt if curricular requirements in the department or university change in the future and allow for developments in the field to become new courses.

To "Create a range of housing opportunities and choices," as smart growth advocates advise, parallels the guideline of "Mix land uses." The requirements for the PTW program should provide a variety of ways for students to live in or inhabit the program. A range of different types of students can be accommodated, including students who transfer into the university or come to the program late in their careers, students for whom PTW is a secondary interest, students who have already obtained a degree but return to the university to take advantage of the program, and students with interests in a number of subfields of PTW that the program can offer while guarding against attempting to cover too much (a constant struggle, we readily admit). While a program cannot satisfy the needs of every potential student, identifying alternate means of approaching PTW and alternate goals for students seeking this instruction will help a program to grow in a manner that maintains flexibility in requirements and maximizes the program's potential.

SELECTING APPROPRIATE TECHNOLOGIES

In selecting appropriate technologies for a system, Fiksel emphasizes that the most recently developed technologies are not always the best and most effective ones and may actually affect the environment in more negative ways than do older technologies. For Fiksel, the best technologies increase the efficiency of the system but also make it more flexible and adaptable. With that in mind, the selection of technologies can also be productively informed by smart growth principles two and six: "Create walkable neighborhoods" and "Mix land uses." Walkable neighborhoods are those that provide safe and easy access to needed goods and services while promoting a sense of community. These sorts of neighborhoods are developed through mixing land uses, incorporating residential, commercial, retail and open spaces into one area. Applying these principles to program growth and development entails selecting a range of technologies to incorporate into PTW courses that both prepare students for work in organizational contexts and foster community, accomplishing multiple goals at once. For example, our program recently purchased on-screen video development software[3] so that students can create software training videos, a skill which is in some demand by private industry in our area. This software can also be used to create a sense of community by enabling students to develop a bank of training videos about how to use other software, such as web page editors and publication design software[4], that can be used for instructional purposes by students in other classes. Participating in producing programmatic resources helps students to feel a sense of accomplishment and belonging, particularly if their contributions are recognized in other venues or courses. Our program is also using technology in the form of an email discussion list to maintain connections with and community among our alumni.

The smart growth principles emphasize that mixing land uses is key to fostering the walkable neighborhoods, spaces that are safe and reasonable to negotiate. To accomplish this, the technologies employed and used within the program should be targeted and overlapping across campus, so that students can feel a sense of building a knowledge base and avoid fragmentation. To this end, faculty within and across programs can hold discussions about the various technologies that they employ and make an effort to learn about and use some common technologies so that students do not have to relearn how to negotiate each course as a foreign land. This sort of faculty sharing may have the happy byproduct of creating a greater sense of community among the often diverse group of part-time and permanent faculty who teach in the program. Because resources for purchasing technologies for computer classrooms are often sparse, it may be necessary to agree as a faculty which applications are key to the program and would benefit the largest number of courses and students and focus efforts in obtaining those. Such programmatic consensus can reduce waste and help make all participants more flexible as teachers and willing to learn new ways to accomplish tasks that focus on helping students learn particular skill sets and critical analysis strategies. We have been more successful at achieving such consensus and sharing such information within our department and program than across departments, partially due to differing goals and resource allocation. For example, we want to begin an electronic portfolio initiative within our program. Upon discovering that the School of Education already had an electronic portfolio requirement for all students, our coordinator met with the chair of that department to investigate sharing expertise and technological resources. While we gained valuable advice as a result of this meeting, we did not choose to employ the technology that the education faculty used largely because it required that each student pay $25 per semester for its use. We did not wish to place such a burden on our students when perfectly viable open source alternatives are available[5]. Ideally, students could rely on the same technology campus-wide to meet portfolio requirements.

DEVELOPING A SYSTEM DESIGN

The design of a resilient system entails the integration of aspects of the system discussed above, including the goals, requirements, technologies, resources, and constraints, to create a viable and navigable system. As Fiksel emphasizes, "Sometimes the greatest resilience is achieved through design simplicity, which reduces the chances of unexpected failure or disruption" (5336). A simple program design would be straightforward, transparent to participants, consistent with previous decisions, and reflective of the faculty and student populations involved. In prompting the system to thrive and then to grow, administrators will do well to follow smart growth principles three, eight, and four: "Encourage community and stakeholder collaboration," "Provide a variety of transportation choices," and "Foster distinctive, attractive communities with a strong sense of place."

"Encourag[ing] community and stakeholder collaboration" is one of the most important smart growth principles and possibly the most difficult principle to apply to program development and growth. As the Sustainable Growth Network indicates, "Citizen participation can be time-consuming, frustrating and expensive, but encouraging community and stakeholder collaboration can lead to creative, speedy resolution of development issues and greater community understanding of the importance of good planning and investment" ("Encourage"). Regarding PTW programs, stakeholders include the program administrators, faculty, and even some university administrators, while other faculty and students make up the community. Collaborating with students may be in some ways easier than collaborating with faculty members in the departmental community for a variety of reasons. The majority of students involved with the PTW writing program beyond the introductory course elected that involvement, have an interest in the program's goals and subject matter, and can provide input and feedback through their courses and through brief, online surveys; in contrast, faculty in the department outside of the program may know very little about PTW, may have no interest in it, or may even believe that it should not be part of their department. This smart growth principle reminds us of the importance of attempting to reach out to the community as a whole and to make a special effort to inform, educate, and enlist the support of even the most resistant colleagues.

Such outreach to colleagues can be accomplished through special means, such as making presentations to the faculty and holding information sessions for faculty and students, and routine means, including talking about the contributions and issues of the program at department and committee meetings and educating colleagues during peer observations of teaching or meetings of departmental reading groups, if those exist. Additionally, qualified colleagues with talents in adjacent subject matter, such as activist writing, travel writing, policy or grant writing, memoirs, and literature of the environment, can be recruited to teach courses that count towards the PTW major requirements. Involving these colleagues as teachers accomplishes two things, both of which could lead to greater long-term investment in the PTW program: 1) these faculty become, for that semester, part of the program, 2) these faculty become familiar with our students and more aware of their accomplishments and the concepts they are learning in other courses in the major.

Another way to keep a broad range of the community informed and active in the PTW program is to seek as diverse an advisory committee as possible by recruiting one or two members from clearly non-PTW fields, so that other voices can speak for the program during department discussions, particularly those concerning resources and hiring. Furthermore, students can also be asked to serve on the PTW advisory committee as can a few providers of internship experiences or local alumni. These individuals may be ex-officio members, and they may not need to attend every meeting or have a voice in every decision, but their presence can improve the diversity and cohesion of the program.

Community participation can also be extended past the confines of the university through internship requirements and with service learning or community-based learning initiatives to strengthen and polish students' educational experience while developing the university/community relationship. In our program, for example, our introductory courses require a service learning project in order to provide students with an opportunity to experience an organic, complex writing situation and to develop a sense of civic responsibility. Through this project, students also discover that they have much to offer the community, and successful projects provide good public relations for the university and the PTW program in the community and on campus. Service learning initiatives also allow us to reach out to stakeholders at the level of university administrators. One of our university's strategic goals includes service learning and community involvement, and making our initiatives known to upper administration helps to demonstrate the contributions of our program to university-wide goals. Furthermore, our university and our college more specifically have a goal of reducing reliance on part-time faculty. Our chair was able to secure our most recent PTW faculty hire by demonstrating how adding this position to our department would allow us lower our reliance on part-time faculty by covering certain courses that part-time faculty commonly teach. We will likely secure our next hire using similar arguments.

The smart growth principle "Provide a variety of transportation choices" serves as a reminder to provide options for system navigation and design to reduce congestion. The flexibility built into the program's design through mixing land use can provide a basis for making it easy to navigate. If the program is configured to facilitate and support double majors, it will be easier to develop routine ways to handle the rules and requirements so that those double majors can progress smoothly. For example, administrators should try to anticipate some of the exceptions to the rules necessitated by the diverse student populations that the program is designed to attract, such as students who complete an internship in another program/department and ask for dual credit, students with below-minimum grades who want to be allowed to continue in the program, students who want a certificate as soon as they complete requirements but before they actually graduate for use in obtaining employment, and students who are less prepared or adept technologically. While it may be impossible to invent specific policies in advance that will cover every potential situation, the program can institute processes to handle situations as they arise, such as course substitution criteria and procedures, and empower a number of faculty in the program to handle these situations so that there is usually someone present to fill out paperwork and give information to students. One way to institute such processes is to draw up charts to represent the delegation of responsibilities among those who will share the tasks of running the program and advising students and discuss these at regular meetings of program faculty. Program administrators can also chart the flow of information that will get students through the program and provide that information in a variety of places, such as on departmental websites and bulletin boards, and to a variety of people, including faculty from outside of the program who are teaching related courses or advising PTW students.

Congestion in the system in the form of inadequate courses to meet students' needs and demands in a particular semester can result from inadequate planning and program oversight. Developing "a variety of transportation choices" in the form of an adequate and diverse number of sections of a required course in each semester, such as a senior capstone seminar, can help to reduce this congestion and avoid trapping students in school for one more semester in order to obtain the courses required for graduation. While our program has been aware of these congestion issues resulting from the rapid growth of our fledging track in the major, we found that without a designated program director, we were less able to coordinate the courses and adequately document the problems faced by students in scheduling classes and graduating on schedule. The need to make long-term plans for the program and designate a faculty member who would be responsible for planning, troubleshooting, and problem-solving gave us part of the justification we required to request a formal coordinator position for our program. While planning does not have to be hierarchical, it does require some degree of coordination and a central person around whom change revolves.

EVALUATING ANTICIPATED PERFORMANCE

Fiksel emphasizes that the evaluation of resilient systems should go beyond outcomes or performance assessment and use predictive measures that help to anticipate the degree to which change and growth in particular directions can be sustained. Some elements involved in such an analysis, which is commonly done through modeling, include economic factors such as operating costs and customer retention, environmental factors such as power use and product reliability, and societal factors such as knowledge enhancement and community trust (Fiksel 5337). As mentioned in the previous section, a clearly identified coordinator and a high level of cooperation between stakeholders and the community are required in order to conduct effective predictive planning as well as outcomes assessment.

Smart growth principles including "Mak[ing] development decisions predictable, fair, and cost effective" and "Foster[ing] distinctive, attractive communities with a strong sense of place" can offer useful guidance in determining criteria for assessing program growth and planning further development. Fair and cost-effective development decisions are those that benefit all participants and community members and expend resources in a just and equitable manner, avoiding short-changing any facets of the community. Likewise, developing a distinctive and attractive community with a strong sense of place involves understanding the goals and values of the community and viewing development as a long-term, iterative process.

Addressing these smart growth principles in program development involves reflecting on and revising the long-term development plans in light of both predictive and performance-based assessments. Program administrators can accomplish predictive growth assessments in a number of ways, such as studying enrollment numbers to judge areas of demand within the program and directing resources there, talking to representatives from other departments about their plans to require PTW courses or even certification for their students, and watching the growth of industries in the region that may employ PTW students upon graduation and determining what skills and experiences might best prepare students for employment therein. Furthermore, administrators should not ignore the desire of some students to go on to graduate school, and the program should be designed to satisfy their needs as well. While specializing may benefit faculty in terms of research, becoming too specialized may not serve the diverse student populations within the PTW program as well as a broad foundation might.

While predictive evaluations can assist in identifying the best areas for sustainable growth, performance assessments can provide useful information regarding the success of current initiatives and provide data to support continuing or intervening in a course sequence, faculty instruction, or technology use. Many universities, including our own, are pushing for outcomes assessment and starting such an initiative can raise the profile of the PTW program on campus. Using a process-oriented assessment tool, such as electronic portfolios of materials collected over time and accompanied by reflective statements, can help administrators to spread the responsibility for teaching students to develop materials over the whole of the program; help students demonstrate the development of their skills, knowledge, and analytical acumen over time; and allow students to see and experience revision on a long-term basis. More importantly, the development of a portfolio requirement prompts a program to codify its goals and values in order to design criteria to use to assess the portfolios and guide their composition, thus perhaps facilitating the emergence of a programmatic identity. As Fiksel and smart growth proponents note, however, this identity is most sustainable when it helps the program to fit well in its environment; therefore, programmatic goals and identity construction should be done with the goals and identity of the department and university communities in mind. For example, if the university values outreach and community involvement, incorporating those into the goals for the program might be useful. Likewise, if the program is housed in a department that values activist or environmental concerns, programmatic goals can also touch on these areas and involvement in or understanding of them might be sought in graduating students' portfolio materials. Smart growth principles emphasize that even outcomes assessment cannot work in a programmatic vacuum; successful and resilient programs reflect university community as well as local programmatic values.

DEVISING A PRACTICAL MEANS FOR SYSTEM DEVELOPMENT

In this phase of development planning, system designers focus on implementing their new developments. As Fiksel explains, stakeholder and community involvement is especially important in this part of the process. As smart growth principle nine admonishes, "Strengthen and direct developments towards existing communities." New initiatives cannot just envision an ideal community population but must measure their effects on existing populations and serve their needs as well. For example, at the start of our PTW program, which is a track in the English major, we accommodated many students who were in the middle of pursuing a degree in English literature. These students often did not take the prerequisite courses, such as the introductory technical writing course, with someone trained in PTW and therefore often required instruction in the basics of writing as a social act or in the use of computer applications that were new to them. Our position as part of the English department requires us to meet the needs of such crossover students in order to make the transition to our new program possible and foster its growth.

Focusing on existing communities also emphasizes building on what is strong in the preexisting environment. For example, as noted above, we as the first two tenure-track PTW specialists arrived in our program to find a strong interest in journalism among students and talented faculty available to teach those courses. Although this focus conflicted with our previous conceptions of what PTW is or should be due to our graduate school preparation, we recognized the importance of developing this aspect of our program because it provides a good foundation and student base for our program, and we both developed courses related to it. Additionally, building on such preexisting strengths helps us to avoid sprawl by trying to take the program in other directions too soon, thereby diluting already sparse human and technological resources.

Perhaps even more importantly than merely following smart growth principle two, "Making development decisions predictable, fair, and cost-effective," in the deployment of programmatic change and growth strategies, is informing stakeholders and community members that you are doing so through documentation of those efforts. Once a program coordinator position has been approved and one has been appointed, it is important for that person to record all administrative duties in order to provide a record and develop data for use in making arguments in favor of creating new positions or acquiring other resources. Some of the activities that might be logged include student contacts; formal and informal meetings with committee members, administrators, and prospective students; time spent in hiring and other staff decisions; completion of requirement checks for certificate students; creation of new courses; and attendance at conferences and workshops to keep skills current. The program administrators should also record the minutes of all committee meetings and post them on the department website or through another semi-public venue in order to create transparency and keep all parties informed of programmatic concerns, developments, and decisions. In addition to record-keeping, it may be useful to hold open meetings of the advisory committee, advertise those meetings, and encourage input from any interested parties. A transparent system, to most onlookers, is a trustworthy one and the more the program wins the trust and cultivates the interest of the community, the more sustainable and resilient it will become.

CONCLUSION

The creation and maintenance of a resilient and sustainable professional and technical writing program asks for a particular emphasis on cooperation and interaction among stakeholders and community members. In addition to this priority, PTW program administrators often face the additional challenge of fostering an inclusive atmosphere in an indifferent or even hostile departmental environment. While professional and technical writing programs seek to find their places within their universities' various departments and structures, we as program administrators can find within systems thinking strategies for linking our work to our larger communities and linking our larger communities to our work. Systems thinkers, stressing the crucial attributes of diversity, efficiency, adaptability, and cohesion, offer us a methodology for building and maintaining stronger programs that serve our constituencies in more and better ways. By "identifying system function and boundaries, establishing requirements, selecting appropriate technologies, developing a system design, evaluating anticipated performance, and devising a practical means for system development" (Fiksel 5330), professional and technical writing program administrators can systematically develop better programs and find new ways to conceptualize problems inherent in existing program structures. In addition, systems thinking privileges the relationships inherent in organizations and environments, the very relationships that can determine whether goals are reached, resources allocated, and initiatives approved.

Using a methodology from systems thinking, we have applied the principles of smart growth urban planning to PTW program administration. Considering programs and their environments as landscapes affords us a way to create and sustain diverse, efficient, adaptable, and cohesive programs. These principles are broad-based and inclusive, fostering collective understanding and cooperation from stakeholders and communities. In addition, smart growth principles, translated for program administration, can help us answer or even avoid altogether the accusation that professional and technical writing programs are the academic equivalent of urban sprawl. Tighter, stronger programs with transparent administration might even mean never again having to hear a long-time colleague ask, "What exactly *is* professional writing, anyway?" And if, by chance, the question were to arise again, smart growth principles and systems thinking strategies would allow us to respond by inviting that colleague to participate in specific ways in our open, inclusive, and mutually beneficial academic community.

NOTES

[1] As of December 2005, the breakdown of English majors was as follows: 112 in professional writing, 91 in literature, 50 in teacher licensure, and 54 undeclared (email from the department chair, December 13, 2005).

[2] Smart Growth America lists the same principles, although in a different order, on their website ("How is Smart Growth Achieved?," 2004).

[3] We first purchased Techsmith's Camtasia, which allows students to make videos of what appears on the computer screen that incorporate sound and other graphic elements. More recently we purchased Macromedia Director for our computer classroom. This software allows students to develop interactive multimedia movies that include other film clips, graphics and audio. We were only able to afford ten copies for classes of twenty, but such sharing can be viewed as a positive way to foster collaboration among students.

[4] We currently use Macromedia Dreamweaver for web design and Adobe InDesign for producing publications.

[5] Drawing upon advice from professional writing faculty at other universities, we are currently investigating the use of the Open Source Portfolio Initiative application, which we will have to house on off-campus server space, as our university's IT department refuses to support installing open source applications on university servers.

WORKS CITED

Allen, Jo. "Compact Planning and Program Development: A New Planning Model for Growing Technical Communication Programs." Models for Strategic Program Development. *Council for Programs in Technical and Scientific Communication Annual Conference Proceedings.* 19-20 Oct. 2000: 56-58. 12 May 2004. Council for Programs in Technical and Scientific Communication Website: http://cptsc.org/conferences/conference2000/program2000/proceedings2000.html

Allen, Jo. "The Case Against Defining Technical Writing." *Journal of Business and Technical Communication* 4.2 (1990): 68-77.

Andrews, Deborah C., & Worley, Rebecca B. "A Networked Approach to Program Growth." *Managing Change and Growth in Technical and Scientific Communication. Council for Programs in Technical and Scientific Communication Annual Conference Proceedings* 11-13 Oct. 2001: 71. 12 May 2004. Council for Programs in Technical and Scientific Communication Website: http://cptsc.org/conferences/proceedings2001/proceedings2001.html

Ben-Zadok, Efraim. "Growing Smart is Hard to Do." *Planning* 69.9 (Oct. 2003): 32-35.

Blakeslee, Ann M. "The Case for an Integrated Approach to Program Development." *Models for Strategic Program Development. Council for Programs in Technical and Scientific Communication Annual Conference Proceedings* 19-20 Oct. 2000: 24-25. 12 May 2004. Council for Programs in Technical and Scientific Communication Website: http://cptsc.org/conferences/conference2000/program2000/proceedings2000.html

Blythe, Stuart. "The Value of Seeking Interdisciplinary Models for Smaller Professional Writing Programs." *Models for Strategic Program Development. Council for Programs in Technical and Scientific Communication Annual Conference Proceedings* 19-20 Oct. 2000: 26. 12 May 2004. Council for Programs in Technical and Scientific Communication Website: http://cptsc.org/conferences/conference2000/program2000/proceedings2000.html

Bosley, Deborah S. "A Proposal for the Marriage of Technical Communication and WAC/WID." *Models for Strategic Program Development. Council for Programs in Technical and Scientific Communication Annual Conference Proceedings* 19-20 Oct. 2000: 72. 12 May 2004. Council for Programs in Technical and Scientific Communication Web site: http://cptsc.org/conferences/conference2000/program2000/proceedings2000.html

Cooper, Marilyn M. "An Ecology of Writing." *College English* 48 (1986): 364-75.

Cooper, Marilyn M. "Foreword." *Ecocomposition: Theoretical and Pedagogical Approaches*. Ed. Christian R. Weisser and Sydney I. Dobrin. New York: SUNY Press, 2001. xi-xviii.

Davis, Marjorie T. "How the Institutional Home Affects a Program." *Models for Strategic Program Development. Council for Programs in Technical and Scientific Communication Annual Conference Proceedings* 19-20 Oct. 2000: 19-20. 12 May 2004. Council for Programs in Technical and Scientific Communication Website: http://cptsc.org/conferences/conference2000/program2000/proceedings2000.html

Di Renzo, Anthony. "The Great Instauration: Restoring Professional and Technical Writing to the Humanities." *Journal of Technical Writing and Communication* 32.1 (2002): 45-57.

Dobrin, David N. "What's Technical about Technical Writing?" *New Essays in Technical and Scientific Communication: Research, Theory, Practice.* Ed. Paul V. Anderson, R. John Brockmann, and Carolyn R. Miller. Farmingdale: Baywood,1983. 227-250.

Dobrin, Sydney I., & Weisser, Christian. *Natural Discourse: Toward Ecocomposition.* New York: SUNY Press, 2002.

Fiksel, Joseph. "Designing Resilient, Sustainable Systems." *Environmental Science and Technology* 37.23 (Dec. 2003): 5330-5340.

Geller, Alyson. "Smart Growth: A Prescription for Livable Cities." *American Journal of Public Health* 93.9 (Sept. 2003): 1410-14.

Hocks, Mary E., Lopez, Elizabeth Sanders, and Grabill, Jeffrey T. "Praxis and institutional architecture: Designing an interdisciplinary professional writing program." 12 May 2004. *academic.writing*1: http://wac.colostate.edu/aw/articles/hocks2000.htm

Lutz, Jean A. & Storms, C. Gilbert. "Introduction." The Practice of Technical and Scientific Communication: Writing in Professional Contexts. Ed. Jean A. Lutz and C. Gilbert Storms. Westport, CT: Ablex Publishing, 1998. vii-xvi.

Rutter, Russell. "History, rhetoric, and humanism: Toward a more comprehensive definition of technical communication." *Journal of Technical Writing and Communication* 21.2 (1991): 133-53.

Smart Growth America. "How is Smart Growth Achieved?" 2004. 12 May 2004. Smart Growth America website: http://smartgrowthamerica.org/sghowto.html

Staley, Samuel R. "Markets, Smart Growth, and the Limits of Policy." *Smarter Growth: Market-Based Strategies for Land-Use Planning in the 21st Century.* Ed. Randall G. Holcombe and Samuel R. Staley. Westport, CT: Greenwood Press, 2001. 201-217.

Sustainable Communities Network. "About Smart Growth." 2004. 24 February 2004. Smart Growth Online: http://www.smartgrowth.org/about/default.asp?res=80.

Sustainable Communities Network. "Encourage Community and Stakeholder Collaboration." 2004. 24 February 2004. Smart Growth Online: http://www.smartgrowth.org/about/principles/principles.asp?prin=10&res=1024

Weiler, Michael, & Pearce, W. Barnett. "Ceremonial Discourse: The Rhetorical Ecology of the Reagan Administration." *Reagan and Public Discourse in America.* Ed. Michael Weiler and W. Barnett Pearce. Tuscaloosa: University of Alabama Press, 1992. 11-42.

6 Curriculum, Genre and Resistance: Revising Identity in a Professional Writing Community

David Franke

When I was invited to direct our Professional Writing major, the first steps were clear: my PTW colleagues and I were to find students, promote the program, and develop a curriculum.[1] Much of this work was informal and occasional—conversations in the elevator with the dean, talk in the mailroom. My colleagues and I were surprised, however, at how much of our time and energy was devoted to writing, and writing that was not exactly scholarly. Our subject-matter expertise played a much smaller role than our rhetorical ability: we learned quickly how to make a complex point simple, what points not to raise, and how to anticipate the niggling unasked questions of our readers. Functional writing, in prescribed genres, was how work got done: getting the program proposal to the bureaucratic center of our system in Albany, NY was a labyrinth in its own right, but then came the course proposal revisions, emails, funding requests, webpages, syllabi, memos, minutes, class-size projections, assignments, and the like, each of which serving as an "important lever" that allowed us to "advance our own interests and shape our meanings in relation to complex social systems" (Charles Bazerman 79).[2] As Bazerman says elsewhere, these genres are not cold and mechanical, but "forms of life, ways of being" ("Life," 19). In other words, in a complex literate system such as a professional writing program, our ways of being—our behavior, our identity, our style—are strongly shaped by the way we engage with key administrative genres. In the pages that follow, I want to tell the story of our program's evolution as embodied and enacted in our administrative writings, and I focus on the curriculum because it is the center of this web of genres. Although the curricular text we wrote is neither profound nor even very long, being nothing but a completely humorless and efficient page full of prerequisites and other technical paraphernalia, it defines the nature of our program and the way subsequent and linked genres are written. Once the curriculum is approved and published in the college catalog, we become animate.

As the center of this "web" of genres, the curriculum is often printed (or downloaded) to a page or two of the college catalog, and it serves as a semi-legal document that gives sequence, shape, unity, themes, and minimums to students, providing them with a loose road map for how they can complete a degree in under four years. Most curricular systems are complex gerrymandered intellectual districts when it comes to course requirements—*a minimum of two from category A and three from category B, but at least six all told,* for instance—and this system serves as the program's DNA, what potentially gives it life and order. Just as we ask people to spell out words when we want to really understand what is said, so too we look at a program's curriculum in order to truly make sense of it. Many conventions are widely accepted: the sequence of courses in the curriculum is indicated by prerequisites and level indications (such as 300-level courses for juniors, etc). The courses, for all their richness, are usually written in deadening bureaucratese, never read until necessary, and perhaps for those reasons the descriptions retain a sense of finality and authority, what Bazerman might refer to as a "reducible" genre (90), one that "exists only in its consequences." And despite the reductive quality, this authority is something that faculty are likely to appreciate, especially after struggling two or three years to get courses through the system and into the catalog. Because curricula are written, they tend, over time, to appear factual, not contingent; purely practical, not theoretical; a firm answer to a set of fixed problems rather than a tacit question about how a program can best adapt and grow.

Yet these conventional assumptions are incomplete. The curriculum, while "reducible," is a form of activity that engages dynamically with the other powerful genres common to a writing program. This case study examines the curriculum not in terms of the logic and technicalities of our graduation sequences and requirements, or even the frustrations of finally getting the thing into print (though doing so *did* severely test our patience), but rather as a source of both continuity and change. In our experience, the curriculum is in fact less like a pronouncement from Zeus than a dialogue with Hermes, both the messenger and trickster, stabilizing and destabilizing our program. By learning to respond to this dynamic, we came to value our functional, administrative writing; in turn, we came to understand better how programs mature and how writing functions for members of a small community such as ours. The effect of our developing understanding and rhetorical savvy is not just that we became better at manipulating the administrative genres of our program—though I think we did—but also that we came to understand better how to sustain a small academic community of "writers-in-training"—a category that includes ourselves. I am advocating that program designers do more than simply "expect the unexpected" or "remain flexible," but rather that they intently look for places to take reasonable risks, and the curriculum is often the most important place in a writing program for that to happen. It is hoped that this narrative will help other program designers decide what a "reasonable" risk might be given their particular situations.

SITE

SUNY Cortland is a semi-rural, mainly tax-funded, solidly established branch of the State University of New York system. The division between the liberal and applied arts is especially sharp. A former "normal" (teacher-prep) school, we still carry the pre-professional major of Education as our largest contingent, followed closely by Recreation and Sports Management; "traditional" Arts and Science majors, those not in a professional track, take up only a third of most incoming freshman classes. Furthermore, many of our students are first-generation academics, perhaps not encouraged by family to entertain seemingly frivolous majors. We are not an endowment-rich school and must therefore work within a very tight and unpredictable state budget. There is little largess for experimentation; it is expected that any venture show a clear and positive relation between expenditures and results – an approach most students are likely to understand well.

For all these reasons, the college, like the culture at large, is pushed to understand success as a lack of error. The number of solecisms in grammar, usage and mechanics can be what determines "good writing." Casual conversation can turn into a lament when the topic of student writing comes up, and too often "students nowadays can't write" emerges as a commonplace marking the travails of teaching. Teaching writing is too often understood as remediation, an unfortunate prerequisite to the real content any course might offer, a way of displaying remembered knowledge, rather than as a process of making or discovering knowledge. Despite a dynamic and persuasive Composition Program and WAC director, writing can function more as an inoculation against diseased prose than a way to join a community and tradition of inquiry.

We developed a writing practice in our PWR program that is often at odds with these conventions, and did so structurally. Our goal, most textually embodied by the curriculum, but echoed in syllabi, assignments, and a thousand other pieces of writing, is to graduate writers who are creative professionals, able to imagine the *textual* needs of their community and immediate audience. Our mantra is that students need "to be taken seriously" as writers, and getting that to happen in our program means they must absorb a rhetorical awareness and familiarity with the conventions of grammar and style, as well as the ability to invent and complete new writing projects. To reach this goal we made our program commodious enough to attend to creativity, analysis, technology, history, theory *and* practical skills—in other words, we chose to build a program that approached professional writing as a liberal art and committed to developing students who will, in order to be successful, have to understand and join writing communities as creative professionals—not merely avoid error.

Our English department's focus on reading and historical periods—no writing courses—allowed us to create, without competing, the several writing strands indicated below. Undergrads must take eighteen credit hours of required courses (with asterisks), and fifteen credit hours of elective professional writing courses, six hours of which must be at the 400 level.

	Creative	Workplace	Rhetorical	Digital	Bookends
200-level	Writing Fiction Writing Poetry			*Writing in the Digital Age	*Introduction to Professional Writing
300-level	Writing Creative Nonfiction Writing Children's Literature Writing Sports Literature	Grant Writing Technical Writing *Revising and Editing Writing for Online Publication Business Writing	*Rhetoric	Writing in Cyberspace	
400-level	Advanced Creative Writing Experiments in Creative Writing	The Publishing Industry *Internship in Professional Writing	The Evolution of Writing	Contemporary Poetics	*Senior Seminar in Professional Writing
grad					

As with many PWR programs in English departments that "have begun recognizing the power of a more eclectic writing program," our challenge is to make coherent course offerings that are united mostly by what they are *not:* literature. The net is cast very wide, from creative writing to "technical and business writing, feature writing, autobiography and biography, research and other modes of advanced composition," (Adams 152). Yet at the same time, our "property" is also spoken for by our extended family in the humanities: journalism, literacy, communications, English, business or management, composition and rhetoric. While most PTW program leaders would be quick to make alliances with some of these before others, the fact remains that we are operating both in the margins and at the intersections of disciplinary territory. It's an odd place and ultimately, I'm not sure it is possible to resolve. As I discuss below, we experimented with several structures to lasso the disparate courses together, but never felt fully satisfied and comfortable. On reflection, I suspect this is simply something we have to accept and I would argue with Adams, above, that this is in fact our strength. We are forced to constantly reflect on our practices and offerings, and there is little room for complacency; likewise, however, it takes a long time in an institution to develop the momentum and recognition that other departments are born to, despite the fact that the courses we teach are informed by a rhetorical lineage that extends back over three millennia.

DESIGNING IDENTITY

Our first curriculum was a loose collection of courses, a list composed by the diligent efforts of faculty who wrote proposals before any of us were hired to the proposed PTW program. Yet we started getting students in our courses even before all of our faculty were hired. And once the three of us were in place, we immediately began writing course proposals, researching other programs, talking to students, contacting potential employers, imagining sequences, picturing our program in disciplinary terms as a place to develop knowledges and practices that would be unique in the context of our pragmatic college. The goal was to create an identity for ourselves, a "space" in which certain kinds of conversations could take place about style, process, rhetoric, and technology. We hoped to actually hear these topics being bantered about in and between classes, to have readings in the afternoons, to connect our program to the ongoing WAC work and faculty development writing in our college.[3] We were trying to design a community, not just a set of classes, and assumed that once we "published" our PWR curriculum in the college catalog, we would be done: our program would be in place, the black and white document would function as a machine to automatically sustain this small academic colony.

STUDENT WRITERS

Students generally joined PWR with an open-mindedness that allowed them to experiment with a broad spectrum of genres and concepts. Even before we finished our first revision of the curriculum, we accumulated our small village of creative, irreverent kids who seemed capable of anything and surprised by very little. Our first line-up of courses was bare: a few creative writing courses, a technology course, a tech writing course. From the start we hoped to elicit reflection, judgment, and a life-long practice of writing and felt we were at odds with the conventions of our college, in that we taught writing as a strategy rather than a skill, linked more to reasoning and imagination than polish—though now in retrospect it's clear that being at odds with the conventions of the *someone* is part of being a new program. Where rhetoric was assumed to be facile posturing, we developed rhetoric into a course on "being taken seriously," and though technology was often assumed to be a set of recondite technical skills, we developed courses that assumed new media to be culture-altering and mind-altering. With the focus on the rhetorical situation, audience, authority, and motive, we were able to move into and among various genres with facility.

Our first true draft of the curriculum was quite broad and, as Kathleen Adams recommends, we tended to teach and talk about writing in a way that blurred the lines between PWR's disparate sources and traditions (152). We sincerely hoped that the disparate motives for writing implied by the curriculum—writing used for play (creative), for solving problems (technical and business), and for critical reflection (rhetorical theory and history)—and through it all our emphasis on technology—would intertwine and fertilize each other. As we added more courses we also assumed that the differences in subject matter between, say, a grantwriting course and a poetry course would become secondary to the strong unity provided by reflection, peer-review, collaboration, audience analysis and revision. We would emphasize close reading of any text, be it technical, creative, or digital; promote an ongoing analysis of motive, content, purpose, and situation; approach grammar as a strategy, not a shibboleth. It would be a struggle, of course, but we aimed to create a set of practices and perspectives that would allow students to speak each other's language regardless of what course they enrolled in. We expected a peaceable kingdom, and waited for our solid and published curriculum to guarantee us just that. While we encouraged students to do some "free range" thinking while in the program, it's also true that their curriculum was pretty strictly managed. As I discuss below, it is perhaps a little ironic that our students were expected to embrace their freedom in the terms we dictated.

A TIE-DYED MAJOR?

Our well-wrought curriculum was not playing out as we had intended in reality. Our students, instead of connecting to "writing" or "rhetoric," broadly conceived, instead attached themselves with a passion to certain genres and formed small sub-groups that codified and confirmed an increasingly restricted writing identity. They played it safe. The larger group, a fairly tie-dyed group of "creative writers," soon took all their classes together, generally eschewing the more theoretical classes, a group that included a trial creative nonfiction class, the digital writing (technology) classes and, most intently, the technical writing class. Though a small minority of techies emerged as the mirror image of these creative writers, most of our students became deeply invested in developing a "voice" and a body of work that could be read aloud at one of the many public performances and poetry slams, both on campus and at nearby Ithaca, New York. Certain poems and stories soon became touchstones for this dominant community: Martin's long poem about coming out, read aloud to his surprised peers at a public gathering, or Tanika's fictional account of an attempted suicide that chilled many readers.

But the development of a shared history was only one manifestation of this group's identity; just as we had defined our program by its contrast with the college at large, these student-writers were defined by what they were *not* writing. I first came to recognize the students' identity-by-contrast when I was teaching technical writing in our second year. We were reading some of the scenarios provided in our technical writing textbook and going nowhere fast. Most of my students couldn't get beyond the immediate personal details of the characters in the scenario who were, if I remember correctly, simply trying to buy forklifts for their company. The emotional / personal interrelationships of these fictional characters seemed to be extremely important to them—my students were obsessed by whether the co-worker might be a slacker or the boss a tyrant—and the writing problem, the challenge to them as technical *writers,* was either misinterpreted or overlooked.

The students' confusion has to be put in context. This is a group of young adults who have successfully negotiated the political and administrative problems of juggling friends, relationships, one or two jobs, a full load of classes and, for some, the demands from home placed on them by their children and families. Identifying and solving problems was not beyond their ken. But it was not until I brought in my own personal issue to class, the need to write an effective response to a major company that sold me a poorly designed hard drive for my computer, along with all the attendant emails and correspondence I had accumulated, were my students able to see the writing as a means of problem solving. When I asked these same technical writing students to write a letter of complaint for me, I received many adequate responses, some equal to my own draft, and several wonderful, excellent examples. Many were expert ventriloquists and did an excellent job of speaking for me in their letters, picking up on my "voice." I am not ashamed to say I cribbed some of their strategies, the ways they positioned themselves as the consumer, delineated the problem, and persuasively argued for a particular solution. I suspected then and still do that these early PWR students wrote from a sense of community—"us" against the forces of coldness and technology—a community that developed its most fluent voice and vivid identity when challenged by "foreign" discourse.

It was clear that we had come to a kind of stasis—a quiet crisis of homogeneity, at least within this large group of creative writers. The majority had become surprisingly self-satisfied with their small constellation of genres. As writers, they didn't seem to be *working* in the ways that we expected. They didn't seem curious or invested in what was "outside" their immediate domain. What encouraged this parochialism? There are the usual suspects: a distaste for "mainstream" academic argument, fear of working hard and failing, the thrill of being able to take the self as a subject—but there were bureaucratic reasons as well. Taking a close look at the way we described the courses, I found that after students got beyond a small set of "core" PWR courses, we only really described two tracks or "clusters": one led into creative writing (Writing Poetry, Writing Fiction, Writing Children's Literature, Experiments in Creative Writing) and one led in the opposite direction to technical and business areas (Computer Technology, Business Writing, etc). There were many shades of gray, but our students seemed to insist on the black and white.

Their resistance was surprising and troubling. Dr. Victoria Boynton, also in Professional Writing, found that her poetry class had several disaffected technical writers in it who were seemingly unable to picture themselves as "readers" of each other's creative work and were having small emergencies of confidence. Dr. Alexander Reid, also in PWR, reported that his new media theory classes seemed to produce anything but a body of enthusiasts for the theoretical and practical issues brought up in his discussions. It was too "cold" to some, too "abstract" and too "impractical" for others. Instead of producing a pervasive program *ethos* for our thirty or so majors and minors, we had unwittingly produced writers who were constantly undergoing minor crises. Small groups were defining their collective selves as being allergic (or immune) to genres outside their purview; for these students, "foreign" genres were threatening and uninteresting. We were not producing writers who were commodious and inclusive. We were producing niche writers who shunned the difficult and unfamiliar. We had written the wrong curriculum.

IN OUR OWN IMAGE

A small program, we three faculty had few options. We could force a broad range of genres on students by increasing the number of required introductory courses, dredge up the truism of how the "real world" expects great flexibility in writers, or just come out and tell students we wanted them to assume a commodious writing identity more in line with our expectations. I am reminded of Richard Bullock's admission of the deep urge to create students in our own image, to have them *"become like me"* (21). And though we wanted students to take up our pluralism, not to wall themselves off to courses that were pragmatic, realistic, and unfamiliar, the irony is that our students were acting just as we were, defining themselves by resisting. But they didn't do it in a way we found comfortable. As teachers, we tended to dismiss our students' "creativity" as shortsighted; yet as program designers we had gone out of our way to de-emphasize the discourse of "writing as correctness." The curriculum we had developed gave both groups—PWR faculty and students—an identity-by-contrast. In fact, we had created students who were, in deep ways, very much like ourselves. This pointed to some difficult questions. In what ways might our program's identity be as narrow-minded as our students? If so, how does community grow past its first identity?

CONSTRUCTIVE CONFLICT

Near the end of one of our first semesters, I stumbled across Chris Anson and L. Lee Forsberg's useful discussion of how writers—in this case, interns in a new environment—created identity. As they put it, "Conflict and initiative seemed to be relatively concurrent in the cycle of transition" (218-219). In other words, Anson and Forsberg found it possible to picture moments of conflict as inevitable, even as a necessary part of development.

We started to look more closely at these small moments of crisis. It seemed that students were doing a very good job of forming a writing identity, which we saw, perhaps more vividly than they, as not only an individual "writing self" but a self-among-others, a Vygotskian social self where meaning was made by the hard-to-see collaborative work produced by students reading and writing texts written for particular social purposes within particular social contexts (*Thought and Language,* 1962). Seen from this framework, their reluctance to change—their deep commitment to one particular image of themselves—could be understood as not so much a personal writing block or distaste for particular genres, but a necessary moment in which identity and meaning are made. It was literally a "pre-liminary," a hesitation at the threshold. Our best goal might be to support, not lament, the way students reactively formed sub-communities of writers, to explicitly identify and seek those places where they could confirm their identity in one area and *from there* explore the big world of alternative writing identities. Instead of seeing a writing identity as a destination—for a program or a student—I came to see it as a necessary but contingent role, an identity-by-contrast, a set of attributes, behaviors and attitudes such as a character from a play or novel might possess.

If I gave up my attempt to define a "professional writer" as someone who seamlessly moves through various rhetorical situations, I gained the ability to see what these writers were actually doing in our program—and what we as faculty were doing. The successful professional writer was perhaps better understood as someone able to join with the struggles for authority and identity in one community, and only after immersion in that community, imagine the conflicts and purposes of other, less familiar situations, again not unlike our attempts to build our program's identity into a small island, separated from the larger academic community. My students, connected to creative writing, had refused to be "managed" by rhetorical theory or fundamental skills of writing; they were ineluctably drawn to celebrate identity as a writer's key accomplishment. We could accept this as their first principle, and only then begin to imagine other writers' identities.

I started to imagine ways to base my classroom questions on identity. What clichés describe a writer of digital media or technical documents? What do such writers really think about their role? What do they really do? What does a short story writer know about organizing in his genre, and what expertise in creating patterns might he bring to the challenges of organizing a business proposal? What does "the writing process" *mean* to a writer in a different situation? When I personalized my problem with the buggy hard drive problem, I had been on track. By emphasizing the lines between writers, the tension between "them" and "us" was highlighted and made more useful, not erased or transcended. Only by attending to these conflicts did my students get what they needed to proceed, a "home base," a perspective from which to eye, with mixed curiosity and suspicion, the new. Creating an identity, however provisional and mutable, needed my attention more than the possibility of teaching my students skills that were immediately "portable."

CONTINGENT IDENTITIES

It was a short step to apply this not only to the individual student writers, experiencing their own crises when confronted with unappealing genres, but to the program as a whole. We as faculty, those who had painstakingly designed the program, were ourselves engaged in a conflict that pitted our aspirations for a "peaceable kingdom" against our students' unexpected resistance. The main document we used to establish our expectations was the curriculum we had written. The curriculum did not only organize our program, giving it shape and character, but as we could now see, also built in complications. Just at the curriculum "shaped" our students (in ways we didn't expect), so they in turn created the exigency—a conflict—that propelled us to take the initiative and reflect on our status. We had rediscovered that even "reducible" documents are endemic. Or to take it home to our situation, we had to learn there is no "identity" cut loose from the complex swirl of texts and communities one writes within, contexts that produce both frustration and the initiative to change. Really (re)committing to our students' development as rhetors meant giving them a room with a view before we asked them to roam the neighborhood. Likewise, the development of our program required that we had to accept the reality of what we were handed: a difficult college context to develop a writing program, a depressed rust-belt employment situation not favorable to writers of any stripe, a small faculty and limited resources.

We soon began three changes. The first and hardest was to recognize that we were ultimately competing with other liberal arts degrees to provide students with an identity. Like it or not, we had to recognize that the diploma was, for many students, an elaborate nametag. Not a job, not a way of life, not a ticket into the Western Tradition. As our students had tacitly asserted, the royal road to their identity was most often, in PWR at least, through creative writing. In PWR we were selling the opportunity for students to recognize themselves as "writers," and we could only set the stage for their future development. In other words, it was time for us to lighten up.

This implied opening up the curriculum again. This time, however, I think we began to see that treating curricula as they really are, as contingent documents, which allow us as administrators and teachers to develop new ideas and make interesting mistakes. We started to see that change was inevitable and necessary. Not only are the curricular requirements always subject to reinterpretation—as second-semester seniors have sometimes taught us—but what the curriculum "spells out" is also always changing as courses develop over time, teachers gathering more experience and learning how their courses are connected or incommensurate. We came to see the curriculum as a key, as in music, in which the program carries on. Further modifications seemed less heretical and more inevitable as we came to see that anything we wrote as a curriculum would set in motion a series of responses—the activities of writing, reading, teaching that occupy and define us a community—and these responses eventually created the need for further adjustments in the curriculum.

It should be noted that these changes, however, are not simple revisions: they required of us that we engage in constructive arguments, read and research other programs, discuss our students' writing and summary evaluations to start. Then a long and excruciating process of creating new course descriptions and courses began, meeting with faculty from other departments, arguing our case in front of various committees—some of which disagreed with each other, making progress seem impossible. Patience, not insight, is what kept us growing and changing, and this process, however uncomfortable, was absolutely necessary to our success.

LOTS OF ROOMS, LOTS OF VIEWS

Developing for our students a room with a view meant deepening their opportunity to establish themselves as a particular "kind" of writer, and in response we began, now five years into the program, the process of creating and herding through committee the new courses that would allow students to align themselves with an identity. Students needed deeper experience in more and narrower areas for all the reasons I've mentioned above, but we also felt the effect of having our first students hit the job market and we were learning from *their* experience how to revise our program according to regional and local employment pressures. No one was knocking on our doors looking for graduates. We could barely find internships for many kids.

Furthermore, I think it slowly became clear to us also that we could never completely prepare our students for any particular writing career: there simply were not enough faculty nor enough hours or even semesters. We decided the small group of core courses would have to suffice; we quit trying to provide all the theory and context for our writers and turned to our strengths. As teachers, we saw that our curriculum and our expertise tacitly cohered into four areas: play and the personal; form-driven writing that engaged in problem-solving; the study and practice of new media; finally, history and rhetorical theory. To return to Forsberg and Anson, the "cycle" of frustration and initiative is a useful metaphor, but it was the faculty, as (curriculum) writers, not students, who took initiative first.

We modified the program's appearance on the page, making room for the various identities students would create, their various provincialisms. To do this, we simply realigned these four territories on paper, calling them "tracks," more explicitly defining various options for our program: a creative writing track, a workplace track, a new media track, and a rhetorical theory track. We retained an introductory and capstone course, along with the internship. We added more hands-on lab time to the two sequential digital media courses, and added a senior-level digital writing course to extend and deepen the track. Creative writing gained Writing Children's Literature, Writing Creative Nonfiction, and Writing Sports Literature. This new shuffle of the deck helped us more easily visualize and advertise our program's options, and we could quickly show our degree offered many niches (of which Creative Writing was the deepest), thus emphasizing distinct spaces one could inhabit–or visit–while an undergraduate. Whether it was this change or whether we just "jelled" at this point, our identity as a program became clearer. It was a thrill to hear in the hallways "after-hours" conversations about writing and reading, and our enrollment jumped to twenty-five majors and about fifteen minors—with a great many students sitting in just to fill an elective.

BEYOND THE CURRICULUM

Until this point, we enjoyed strong support from our president, dean and chair. Soon, however, the inevitable changeover took place. In a short time we found ourselves with a new Chair, Dean, Provost, and President, not all of whom saw Professional Writing as an integral part of the college's development. We were disappointed in house when English literature faculty (now referred to as "the liberal arts" faculty by our chair) decided to stop counting PWR courses as a legitimate part of the English major's requirements. It was uncomfortable, perhaps inevitable, and unfortunate—many of the promises that had been made when we were hired were now lost in the seas of institutional memory. But we were working seriously. All of us had taught four new courses a semester, kept learning new software, met weekly to plan, and kept up our own writing. The paper load was enormous both from teaching (all our courses were Writing Intensive, of course) and from pushing proposals through the various committees. In purely practical terms, we realized we could not sustain our work at this pace forever. The belief that we could continually create new course options and new combinations of classes was becoming untenable. We had other projects, too: our own creative and scholarly writing, the dream of an MA program in Rhetoric that would let us (for the first time) to teach graduate courses in our own area, the possibility of joining the National Writing Project. There is a human element that needed attention, too, for we were juggling the demands of new families, children, and elderly parents.

At seven years, as the first sabbaticals came into view and we were granted tenure, it was time to review and reassess our first years. Our students had changed from our first tie-dyed contingent. A new community had formed, one that didn't seem to need us to direct them as much; students arrived at our program with clearer ideas about what they wanted to do for themselves. We saw them use writing to pursue their passion, not just to "hone their skills"—a term we never liked. The change was slow but definite. Becki had been fascinated with the environmental and social role that zoos played—and she loved the animals. Because she had no aptitude for zoology or medicine, she used writing as a way to get her foot in the door and soon started writing publicity for the local zoo. Likewise, Raymond, a skilled auto mechanic, decided to join our program so he could pursue his passion for cars by writing better repair manuals than now exist. Others majors joined to work in comics, or to prepare themselves for working in politics; still others went on to become teachers or to attend graduate school in creative writing. We were not a "professional" track in the traditional sense, as was the case for our neighbors in the departments of recreation or education, but we were finding our own rhythm and playing to our strengths, much as our students were doing as writers.

We saw that few of our students were trying to apply the PWR degree to get an immediate job as a freelancer, editor, or technical writer. Our best were going on to graduate school in creative writing or applying to Masters of Arts in Teaching programs. Unlike Recreation or Education majors, for us there was no large institution looking to hire writers; the local rust-belt economy was tightened to the last notch. To develop the maturity and facility needed to move from rural New York to where the jobs are, on the coasts and big cities, would take more than a long time—it might take generations. It was a little unrealistic to say the least to assume that by tweaking our curriculum and pushing students to travel afar for their internships we could meliorate the challenges presented to us by our uncertain students, our local economy, and our new administration.

Yet we could not ignore that our students were enthusiastic about our program, and that it was still growing, presenting us with new problems as other departments asked us to offer service courses for their students, many of whom were anything but expert writers. Technical, business and creative writing were in high demand, but soon all of the courses were filled, from Creative Nonfiction to Writing Children's Literature. Clearly, we had lined up an attractive roster of offerings, but we no longer had the teachers we needed to take the classes. The fourth faculty line we had been promised was clearly never going to materialize, though our classes were more in demand than ever. Our exit interviews indicated that students wanted to be challenged intellectually, to go deeply into a subject, and to have more freedom to pursue their interests. The result is that we were pulled in two directions: on one hand we saw ourselves becoming a service department for other disciplines; on the other hand, we felt we needed to open up to accommodate our students. After all, the best ones weren't leaving us for jobs—they left for more advanced academic work.

We had to recognize that much of what students were learning was happening outside the classroom. Every semester we took students on a writing weekend to a verdant (or gelid) island in the nearby Adirondacks for workshops and readings; the literary magazines had been revived in both print and web forms; our learning communities were taking off; our online international news journal *NeoVox*, through the tireless work of Alex Reid and Lorraine Berry, was serving as our own in-house site for internships. Reid also put our program at the front of the technological initiative from Apple called iTunes University. Students were learning to write by writing, and their audience was not simply the teacher.

Furthermore, as faculty we came to understand better how to see our own workplace writing, seeing that writing (and revising) curriculum was a form of composition, no less challenging or influential than writing scholarship, and in some cases more so. Though we never explicated the "administration as scholarship" argument as developed by Christine Hult nor leaned on Ernest Boyer's redefinition of scholarship—we didn't expect our various committees would be receptive—we had accomplished some good things through the construction and reconstruction of the curriculum over this period. We saw how the functional, administrative drafts challenged us to revise our understanding of how writing governs a community. We started to see the curriculum as a constantly negotiated response to what various communities of students were doing—rather than a set of rules that codified their identity. We came to appreciate and respond to the way the curriculum set in motion certain ways of acting, having direct and indirect effects on how we acted as a community of students and teachers. Our community and its texts developed a sort of feedback dynamic I want to call a "voice" or stance, a certain tone or characteristic way of acting, of asking questions and making decisions. The character of our program, its evolving identity, had been created in large part as a function of how this curriculum resonated with other documents—syllabi, assignments and even student papers. The bureaucratic process of writing our program's curriculum helped us become better writers and better teachers of writing, which in turn shaped our next rendition of the curriculum. I think we became more realistic and even a little more humble. We had to learn to read our curriculum for what it always was: a powerful proposition, a set of propositions about learning that enabled and constrained—not an identity in itself. It was time to use the curriculum as a space in which the program and the students could determine their identities.

THE FUTURE

Our identity as faculty started to change, too. We were able to let curriculum become less of a mirror of our thinking and hopes. We saw ourselves in other projects to pursue: our own creative writing and scholarship, a dream of an MA program in rhetoric, a certificate program in writing, a National Writing Project. We began to look for ways to revise the curriculum to support our strengths. An honest assessment recognized that to some degree we *were* a service program. The courses Revising and Editing, Technical Writing and Business Writing continued to be in high demand by other departments, and courses such as Writing Sports Literature and Writing Children's Literature were a perfect match for the needs of our populous neighbors Recreation and Education. Creative Writing was *always* full, and we were spread thin teaching these classes. We had the good fortune of having excellent adjunct faculty who volunteered to take many of these courses. We were in a secure place. We had a coherent and popular program, excellent faculty and a strong community of students. It was time for one more change to the curriculum.

At this writing, we are again in the thick of revision. Our changes will do two things: first, create courses that build on the work we are already doing. Some examples: a proposal has been submitted that gives students credit for semester-long work that culminates in the writing retreat; another course put into the pipeline rewards students for their public performance of work; a service-learning course has been proposed that will contribute to the community and draw strength and resources from various in-house programs already in place. We're popular, and we recently reduced the number of required courses and increased the electives. Several courses became designated as "general education" courses, thus filling a requirement for many undergraduates. They are now almost always full. This is certainly a long way from the tightly structured program we developed when we began. Student writers can experiment more and, we hope, find their particular "room with a view" as they near graduation. This openness is balanced by an increase in the total number of advanced courses we require, though students again choose exactly which ones. Advanced Creative Writing, for example, will give students a chance to specialize. 500-level courses will entice them, we hope, to stick around for a proposed certificate in writing, and some of the courses from our newly approved National Writing Project site will bring teachers-who-write into our program.

We are obviously in process. We hope that those who, like us, juggle the various hats one wears while designing a program–those of teacher, scholar, administrator—will see in this narrative a developmental arc that speaks to their own curricular work. We have learned to be patient while people figure out where their abilities and passions lie, and that applies equally to our students as to ourselves. The curriculum we struggled to perfect is a powerful tool, the most visible example of our personal, intellectual and pedagogical agendas, but itself only part of a larger system of writing that stretches from short memos to syllabi to ponderous state mandates. While it can trace out a history for a student (and a program), it is an enabling constraint on what is possible. The good judgment that enables one to change (or resist change) is something that can't be published in the college catalog or imposed by fiat. We hope, however, that good judgment is what we have exercised in our revisions over the last few years, and that the resulting curriculum enables our students to learn the same for themselves.

NOTES

[1] I wish to thank my colleagues and friends Drs. Victoria Boynton and Alexander Reid for the intelligence and creativity they shared while we developed this program together.

[2] I found *Genre and the New Rhetoric* edited by Aviva Freedman and Peter Medway (1994) and *Genre and Writing* edited by Wendy Bishop and Hans Ostrom (1997) excellent ways in to the growing sub-discipline of genre studies.

[3] The nascent "Faculty Writing Group" began meeting regularly during this time as a way to bring together faculty to discuss their ongoing creative and academic writing projects. I discuss organizing this group in "Completing the Circle," an article available at http://dinosaur.cortland.edu/facultywritinggroup.pdf

WORKS CITED

Adams, Kathleen. "A History of Professional Writing Instruction in American Colleges: Years of Acceptance, Growth, and Doubt." SMU Studies in Composition and Rhetoric Series. College Station, TX: Texas A&M University Press, 1995.

Anson, Chris and L. Lee Forsberg. "Moving Beyond the Academic Community: Transitional Stages in Professional Writing." *Written Communication* 7.2, (April 1990): 200-231.

Apple Computer: iTunes University. http://www.apple.com/education/products/ipod/itunes_u.html.

Bazerman, Charles. "Systems of Genre and the Enactment of Social Intentions." *Genre and the New Rhetoric.* Ed. Freedman and Medway. Taylor and Francis, 1994. 79-101.

Bazerman, Charles. "The Life of Genre, the Life in the Classroom." *Genre and Writing.* Ed. W. Bishop and H. Ostrom. Boynton/Cook, 1997. 19-26.

Boyer, Ernest. *Scholarship Reconsidered:* Priorities of the Professoriate. Carnegie Foundation for the Advancement for Teaching, 1997.

Bullock, Richard. "Theorizing Difference and Negotiating Differends: (Un)naming Writing Programs' Many Complexities and Strengths." *Resituating Writing: Constructing and Administering Writing Programs.* Ed. Joseph Janangelo and Kristine Hansen. Portsmouth, NH: Boynton/Cook, 1995. 3-22.

Hult, Christine. "The Scholarship of Administration." In Janangelo, Joseph; Kristine Hansen (Eds.), Resituating Writing: Constructing and Administering Writing Programs. Portsmouth, NH: Boynton/Cook, 1995. 119-131

NeoVox. www.neovox.cortland.edu

Vygotsky, Lev Semenovich. *Thought and Language.* Ed. and trans. Eugenia Hanfmann; Gertrude Vakar [Myshlenie i rech', Moscow, 1956]. Cambridge, MA: MIT Press, 1962.

7 Composing and Revising the Professional Writing Program at Ohio Northern University: A Case Study

Jonathan Pitts

Since 2000, when I was hired as an assistant professor to design and coordinate the professional writing program at Ohio Northern University—a small (3300 students) university comprised of a college of arts and sciences, a law school, a pharmacy school, a college of business administration, and an engineering school—the design, coordination, and administration of the program has gone smoothly and well. As a department of teaching generalists, we have nothing but success to talk about. The English major has, over the past ten years, grown from twenty-five majors and minors to just under a hundred, has added two new faculty and replaced two retired, has completed major renovations and upgrades of its offices and classrooms, and has added the professional writing track. Two years ago, the administration approved the hire of a second professional writing faculty member. Two years ago also, the number of English majors (and university enrollment generally) was increasing so fast that the faculty wondered, in a meeting with the university president, if we should try to limit it. Since then, the number of professional writing majors has leveled off (tied with journalism for the second-highest number of majors, behind Language Arts education and ahead of literature and creative writing), but it remains evidence, we like to believe, that the professional writing track has a solid place in our undergraduate English major.

Another number that means a lot to us is the amount of our recent graduates who are getting jobs or admission to graduate schools. Judging by the number of PW program graduates who are doing what they want to do, the program is a success. For me, this means that the program is preparing students well to enter the world of professional writing. The emails I receive from my students about their jobs and their graduate programs provide important material for me to use in selling the program to prospective students and majors—the program does what it says it will do. I describe three of these recent graduates below.

But while such student success is in one sense the most important to us, the English department, and the university, it isn't, and shouldn't be, the only barometer by which to judge the performance of the PW program. As all university curricula must do, the PW program at Ohio Northern must succeed at multiple levels in order to be seen by the administration and the university as "viable." Catherine Latterell is generally correct in her observation that the contexts of PTW programs at small teaching institutions can be characterized by interdisciplinarity, an emphasis on writing, and a comparatively close relationship between administrators and faculty. The purpose of this article is to describe the composition and ongoing revision of an undergraduate PW program at a small university that prides itself on providing students with a liberal arts education with a pre-professional emphasis. I also hope to offer a detailed and perhaps representative example within Latterell's more general scheme.

In this regard, the "pre-professional emphasis" of my school might distinguish the development of our PW program from those at other small teaching institutions. I realized early in the process that I wouldn't have to do much selling of the program to the university, since ONU has for years been committed to the professionalization of liberal education. The idea of an English major devoted to the professions made instant, even compelling, sense to faculty and administrators, so that often it seemed (and still seems) as if the professionalization of English studies was an argument to the professionalized university for the continuing relevance in these times of the English major. And yet, even as the program enjoys the support of a small university, I found throughout the process that our students were best served if the professional writing major was conceived and marketed not as merely a professionalized version of the English major but as a body of theory and practice inherent in the study of English and its responses to cultural and technological change. In other words, at my university there was a nice fit between the disciplinary origins of a new professional writing major and the desire of those outside the discipline (students, faculty, administrators) that English respond to the exigencies of the job market. In this way, I would like the example of my program to be seen as both limited in context but also as a general claim for the genuine value—"viability"—of professional writing programs at small teaching institutions such as mine.

PLANNING AND CURRICULUM DESIGN

The Five-Year Plan

I was hired by Ohio Northern University in 2000 to define and formalize the two-year-old professional writing major. I'd been told that in the two years the major was coordinated by my predecessor, it was a collection of nonfiction writing courses common to English departments—nonfiction writing, magazine writing, newspaper writing, prelaw writing (Appendix A). The department had wanted to gather these courses together into a coherent major after researching similar curricular initiatives at other departments around the country. The Dean of the College was excited about a writing program with a traditional literature and language core integrating the demands of the "real world" job market English majors faced upon graduation. The Dean had pledged to back the development of the new major (and would continue to support the program until his retirement in 2003) and the hiring of a faculty member, my predecessor, who would leave for another job after two years. "All we really know is that we don't want a technical writing program, you know, writing about corn harvesters and so on," a senior English faculty member had told me at my campus interview. "Beyond that, we'd like you to figure out the major."

Like the university itself, the once-small department had exploded in size in the space of a few years, growing from a total of twenty-five majors in 1993 to more than ninety in 2000, the year I arrived. To further define and market its course offerings to accommodate growing student interest in the English major, the department, as many departments did in the eighties and nineties, developed "tracks" of concentration—Literature, Creative Writing, Language Arts Education, Journalism, and Professional Writing—linked by a common core of British and American literature surveys, literature electives, an introduction to English studies, and a course in linguistics and the history of the English language. "We're doing the same things we've always done," said a senior colleague, "we're just marketing ourselves differently. It's a different world."

He was right, for the most part. No new courses had been added, the old courses were simply being presented in new ways. But to represent these courses is to change them, significantly. The "tracks" approach lays a sheen of professionalism over the competent generalism of English, suggesting to students that there are distinct "jobs" out there for "literature" people or "creative writers." There is no reason why a straight literature major can't also be an excellent creative writer or professional writer. But as the most recent addition, the professional writing track implicitly promised students what the other tracks could not: more vocational preparation for, and thus access to, jobs.

But we aren't a polytech, so that while we might offer a course in technical writing or business writing, prelaw writing and desktop publishing, the PW major had to retain a liberal arts background and a relationship to traditional English studies. As Anthony Di Renzo has pointed out, PTW programs housed in the Humanities should reflect liberal arts values (50). At a small university, PTW programs pretty much have no other choice.

As I thought about the PW curriculum and planned its implementation, several unique qualities of the program arose naturally from the immediate institutional and academic context. First, I recognized that the PW program was intended to function as the English department's contribution to the interdisciplinary side of academic life at Ohio Northern. The department's teaching of a three-course writing and literature sequence had long been valued on campus as the foundation of the general education curriculum. But with the exception of new Honors program seminars, there seemed to be little collaboration between departments and colleges on courses. With the secondary study requirements for the PW major, the English department had wanted to develop such interdepartmental relationships.

Interdisciplinary collaborations are, of course, difficult to create and sustain, especially at a small university where departments and faculty with heavy teaching loads and limited resources struggle to maintain their own programs. When I thought about the necessity of eliminating the public relations and art requirements from the existing PW requirements, I wondered if maintaining the department's relations with Communication Arts and Art was more important than my curricular ideas. And yet, as it existed, the PW major had no central narrative, no coherent disciplinary foundation. I argued with myself that perhaps all the major really needed was some meta-course under which its existing courses would make "professional" sense.

The English department had yet to develop courses in media theory and criticism, cultural studies, information design, or digital culture. There was, in fact, no course in media theory in any department. In thinking about the composition of the PW program in the English department, I was, of course, also composing myself—thinking intensely about my own identity and role in the department. My doctoral concentration had been in Cultural Studies and American literature, and I'd had extensive creative, nonfiction, and technical writing experience. I knew I wanted my PW students to see themselves as "symbolic-analytic" workers, as Greg Wilson sees himself and his students and as I'd seen myself for most of my working life. Indeed, if I was to be a professional model for my students, the theme of my working life would be "change," often preceded by the word "bewildering." In one sense, I was your average postmodern information worker; in another, more positive sense, I was a writer, not a worker. The purpose of the PW program at Ohio Northern, it seemed to me in the course of my first year, was to help students understand themselves as I had come to understand myself, as both worker and writer, as postmodern subject and marketable agent.

I saw the program, then, as two-pronged: one side of the major would be anchored by a cultural studies course, the meta-course counterpart to the traditional English course in Literary Criticism. The course would introduce students to cultural and media theory, providing students with a historical and theoretical context for their symbolic-analytic development. The other side of the major would be anchored by Rhetorical Theory, an advanced seminar taking students from classical rhetoric to cyberspace. These two upper-division seminars, Cultural Studies and Rhetorical Theory, would be advanced explorations of topics surveyed in the sophomore-level introduction to professional writing, Writing in the Public Sphere.

Despite our growth, we remain a small department, with eleven full-time faculty. The recent hire of Dr. Paul Bender, a specialist in rhetoric and digital communication, was approved by the administration for a variety of reasons, one of them being the growth of the professional writing major. Dr. Bender was hired to bridge the professional writing and journalism majors, both of which consisted of one full-time faculty member. His hire and our latest revisions of the PW curriculum constitute the final phase in the five-year plan for the development of the program that I put together in the beginning of my first year, in 2000. These course revisions are to take effect in the fall of 2005.

IMPLEMENTATION

It wasn't simply a matter of adding and deleting courses, my chairperson reminded me. Nearly every change we make in our curriculums affects the course offerings of other departments. In removing PR Writing from the PW major, for example, I was de-populating a Comm Arts course. In adding Desktop Publishing, I was replacing a required two-course sequence in Graphic Design taught by the art department. Since Desktop Publishing would also serve Journalism majors, I and the journalism professor, Dr. Bill O'Connell, met with art department faculty and faculty from Comm Arts to explain the revision and to discuss its implications for their departments. Again, I am fortunate to work at a small university, where relations between departments, especially those in the arts and humanities, remain constructive.

The first year I taught Cultural Studies I focused the course on popular culture theory and criticism from the Frankfurt School to the present. Many of my majors were already writing cultural criticism for the *The Northern Review,* the student newspaper, and many were double majors in professional writing and sociology, or political science, or history, where they were also exposed to some popular culture theory. It seemed to me that the course facilitated the cross-disciplinary thinking and writing that was a central feature of the PW major, with its secondary study requirement. In making my case to my department for the addition of the course, I argued that many of my students wanted careers in editing and publishing, and that a sophisticated cultural and meta-cultural literacy were necessary for such jobs (and for the students' work on the PW web magazine, *Delirium*).

Some colleagues wondered about the seeming theoretical orientation of the major—what was happening to the distinguishing "hands-on" quality of the major? With Bill O'Connell's help, I argued that Desktop Publishing, which replaced a two-course sequence in Graphic Design taught by the Art department, offered students the practical skills to balance the theory they encountered in Cultural Studies. Indeed, we argued, the two courses worked hand-in-hand: students couldn't really understand the publishing software (*Dreamweaver, Photoshop, QuarkXpress*) they'd use in the course without some knowledge of the cultural and technological contexts in which such software is used.

In the spring of 2001, as I prepared to develop these two new courses over the summer (I received a grant from the Dean to take summer courses in the three software applications) in order to teach them the following fall, the department of technology was instituting a digital design minor open to non-technology majors. In the technology minor, students took a sequence of web design courses at a level of detail I could not cover in the two weeks I had to devote to web design in the new desktop publishing course. Because our academic quarters are nine weeks long, students had an absurdly small amount of time to spend learning very complicated programs. In offering Desktop Publishing, was I offering "competency," or something much less, and was this okay? Bill O'Connell and I decided it was okay—we were claiming only to introduce our students to the software, not that we would make our students experts.

PROGRAM MISSION

In the fall of 2001 I prepared a presentation on the PW program for Explore the Colleges Day in October, when prospective majors and their parents could find out more about possible majors. Rather than focusing the presentation on the kinds of careers PW majors might pursue (since those are the same careers open to all English majors), I planned to speak about the courses in the major and how they differed from the traditional English literature courses. Then came September 11th.

The September 11th tragedy presented a bewildering range of issues and problems for all of us, students and faculty. And like faculty everywhere, we've worked ever since to provide students with the skills and tools to deal with these issues and problems. The distinguishing feature of the PW program, its focus on mass media and cyberspace, was suddenly, perhaps, urgently, relevant to my students' lives. (An English literature major, Nathan, who would take my Cultural Studies course, later that year wrote his senior thesis on the use of media by terrorists). The events of 9/11 didn't change the concept or direction of the PW program so much as reinforcing the existing rationale for the cultural and rhetorical focus. It was as valuable for my students to learn to think about mass media and cyberspace as it was for them to spend time on technical writing problems. But how would I explain all this to parents and prospective students, who quite rightly wanted to know what training they would get for their money?

That fall we were putting together the first issue of *Delirium*. By this time I had gathered a staff of seven students, and we were preparing to publish a book-length memoir by an Israeli writer, Ephraim Glaser, about his experiences escaping the Nazis. We were also publishing an interview with a local survivor of the Holocaust. We were busy designing, editing and hyperlinking the texts when the planes hit the towers.

Discussing the upcoming issue at a staff meeting, we decided that we would title the issue "Technology and Historical Memory," featuring the memoir, the interview, and an article on Osama Bin Laden and theocracy. By the time Explore the Colleges Day arrived, we were online with a publication that was both visually and textually arresting. I used the webzine to show parents and students what professional writers might do—they could produce our world.

What kind of world would they make? I asked parents and students at my presentation. It's a familiar question, a staple of English department orientations and university commencement addresses. But the question has become more meaningful for English departments in particular, given the centrality of images in the postmodern world. English majors no longer simply create and interpret the texts that give meaning to the world; they (can) create their very reality. In the Cultural Studies course that winter, we studied the images of 9/11, read across the recent history of image and media theory from Benjamin to Debord to Benjamin Barber's Jihad vs. McWorld thesis. Three years later, it was gratifying to listen to a graduating PW senior (Nathan) chuckle knowingly at a recent Yahoo! headline announcing that digital photography and Photoshop had altered our concepts of truth and reality.

I wish all PW majors could graduate with that kind of critical sophistication and facility—and interest in—the urgent connection between the use of technology in the real world and the philosophical consequences of its use. The mission of the PW program as it has developed is to inculcate in students a level of comfort with change—professional, technological, cultural. I explained to prospective majors and their parents that professional writing is both a career and an attitude, a goal and a method of reaching that goal.

The first cohort of the PW program were now writing their senior theses. What was a professional writing thesis? Anything, really, as long the project involved intensive reading, research, writing, and some degree of rhetorical analysis or genre exploration. Steve's thesis involved a rhetorical (textual and visual) analysis of minor-league baseball team websites which he presented using PowerPoint. Allen, a PW major headed to Duke University law school in the fall, wrote his senior thesis on a topic (the effects on Third World economies of U.S. economic aid) he had addressed as an intern the previous summer at a Washington, D.C. think tank. Tracy, a PW/Chemistry double major, wrote her thesis on the role of serendipity in major scientific discoveries; Naomi, wrote (and courageously presented to the public) an unsentimental memoir of her spiritual journey in the decade following the drowning death of her father.

At the presentation of her thesis to faculty, students, and parents, Naomi broke down numerous times throughout her reading, struggling to read through her tears. This made the audience uncomfortable, in particular, two senior colleagues who objected to the overly personal, "unprofessional," nature of Naomi's senior essay topic. In the days following the presentation, I argued for the personal and academic value of Naomi's achievement and for its direct connection to her professional future.

Naomi's topic had originated in my Nonfiction Writing course, which is required of PW majors. Since the journalism major already offers a Literary Journalism course, for the past four years I've taught Nonfiction Writing as an introduction to travel writing, the personal essay, and the memoir. In the near future I might change the course to include business writing, technical writing, and public relations writing, but for now the purpose of the course is to give students advanced experience in the writing of extended narratives for a general audience.

In focusing the course on the memoir, I believe I'm maintaining a tie of the PW major to the central, distinguishing purpose of the humanities: self-knowledge. There will be very few occasions in my students' lives when they will be both encouraged and assisted in writing about themselves. Of course, nine weeks is not enough time to gain substantial insight into one's life, so I make the goal of the memoir (if that's the option the student chooses) a dramatic narrative that might be developed beyond the course. When one of my senior colleagues asked me what the connection was between writing a memoir and a career as a professional writer, I explained the professional value for students of the ability to both narrate their lives and to conceive of their lives as narratives. In fact, perhaps the most important lesson I'd like my students to learn from the course concerns the narrative quality and thematic richness of their own lives. I'd like all of my students to be able to say, as Allen did at the end of the course, "I'm seeing stories everywhere now." Such a lesson is a crucial one for the journalist, the editor, the freelance writer, the media consultant, and to a great extent the business and technical writer, to learn.

Finally, the focus on the memoir in Nonfiction Writing dovetails with the major project in the Cultural Studies course, an experimental, autobiographical website produced by each student from theoretical/conceptual guidelines offered by Gregory Ulmer in his brilliant book, *Internet Invention,* which we use with Bolter and Grusin's *Remediation: Understanding New Media.*

REVISIONS

The latest PW program revisions reflect the final phase of the five-year plan. The formerly required Advanced Writing course has become an elective. The formerly required Magazine Writing course has been replaced with Writing in the Public Sphere. A required Rhetorical Theory course, taught by Paul Bender, has been added, and Writing Cyberspace has been added as an elective. With Bill O'Connell, Paul and I had discussed deleting Desktop Publishing. Students had begun to feel that nine weeks simply wasn't enough time to learn three applications. But most of the students we spoke with liked the class and found it useful, so the course remains. Our recent graduates working with technology in their new jobs continue to assure us that they are learning what they need to know as they go.

Magazine Writing had long been taught as a course in freelance writing, yet so few of the students taking the course were actually interested in freelance writing (I have yet to meet a student who is!), we thought we might make better use of those credit hours. We decided to fold it into the more general Writing in the Public Sphere course. The problem for us was that many of the students taking Magazine Writing were Comm Arts, Public Relations, or Language Arts majors who needed the course as an elective. Fortunately, those departments were amenable to the change. Writing in the Public Sphere now functions as an introduction to professional writing with an emphasis on civic argumentation and critical thinking (course text: Donald Lazere's *Reading and Writing for Civic Literacy*).

The final revision is the addition of a required Grammar in Context course. Most of the English department believes that such a course is sorely needed, and according to informal surveys, so do our students. As might be the case in other departments, most of us were trained to teach process-centered writing; some of us, especially the younger faculty, were encouraged in our graduate school training to avoid teaching grammar altogether. Perhaps the pendulum swung too far in that direction; in any case, we believe once again in grammar, even if, as we know, studies show grammar lessons have little effect on student writing. But the goal of the course isn't necessarily the immediate improvement of writing. With knowledge of grammar concepts, and even simply being able to name parts of speech, students can lay the groundwork for a lifetime of work on their and others' writing. It seems to me now a little shocking that an English major could graduate without having taken a grammar course, though I wouldn't have thought so ten years ago. For future professional writers, the course is even more important.

A final course addition is actually only the first-time use of a course that had already been in the curriculum but never activated—the Professional Writing workshop, a 1-6 credit course on selected writing topics. I wanted to use the course to bring in working professional writers, preferably successful alumni. In the spring of 2002 I'd invited two English alumni who'd graduated before the existence of the PW program to speak to our majors. One alumnus worked as a senior documentation manager for Microsoft, and the other was editor-in-chief of a national magazine in San Francisco. The talks were informative and inspiring. The following fall I arranged for another alumnus who ran his own publishing business, to speak to students about his work. He agreed to return in the spring of 2004 to teach a popular week-long workshop course on the publishing business.

RESULTS: THREE RECENT GRADUATES

Naomi

I recently received an email from Naomi telling me that after a year of looking she'd found her dream job with a publishing house in a Midwestern city. She was starting as an entry-level copy editor and had already received freelance editing work.

As a freshman, Naomi knew only that she wanted to write and edit for a living. I told her that she'd come to the right place—our PW program, of course—and that we'd help her prepare for the career she wanted. She was not a particularly strong writer, but she was an excellent student. As she progressed through her writing courses, she discovered that while she had no serious interest in either the purely theoretical, more academic side of professional writing or in the solely practical and technical, she was passionate about the middle, writing and editing at the juncture of theory and practice. As the founding editor of the PW program webzine, *Delirium,* she was passionate about editing, managing a magazine, and desktop publishing. It was her work on *Delirium* that led to her internship, editing and preparing for publication in *Delirium* the memoir of the Israeli sculptor and Holocaust survivor Ephraim Glaser. Her internship convinced her that she had chosen the right major.

Naomi's internship was successful for all concerned. As they are in most PTW programs, our internships are collaborations between faculty and students. Our PW students are responsible for finding the internship they want, so we are flexible as to what constitutes an internship. The only requirement is that writing be an important aspect of the experience. When Mr. Glaser submitted his book-length memoir to *Delirium,* Naomi and I decided that her internship would consist of editing the book with Mr. Glaser, and of directing its electronic publication. Naomi and I met once a week for a quarter, when we'd work through the week's editing of the memoir, and discuss aspects of editing practice. Naomi found she loved working with authors on their work. She also became interested in writing her own memoir.

In the process of editing Ephraim Glaser's memoir and of writing her own, Naomi found herself making use of some the cultural theory she'd been introduced to in Cultural Studies. In preparing Glaser's memoir for publication, she had to confront the nature of the memoir in the digital age—few people would be willing to sit at a computer to read a lengthy memoir. Novels had already digitally adapted through hypertext; would hypertext work for the memoir as well? Such questions inspired and energized Naomi—they were problems she could seek to solve in her professional life. But they were also personal, intellectual, and professional questions I could not have foreseen in designing and administering the PW program. But working with Naomi has taught me to see the PW program as a set of student resources rather than as a one-size-fits-all training program. And this is appropriate for a small, humanities-based PW major, where faculty have time to work with students on their preparation and their career goals.

Naomi has since begun work on a book project, a historical study of a pioneer family in her Indiana town, and continues to write and publish magazine articles.

Steve

In the fall of his sophomore year, Steve came to me and said he wanted to work in public relations for a professional baseball team. I urged him to consider double-majoring in PW and Technology. He declared the double-major, but after a quarter of computer programming courses, decided to drop Technology and to replace it with a minor in business and public relations. He found that his interest in computers extended no further than the software I was teaching in the new Desktop Publishing and Design course—*Dreamweaver, QuarkXPress*, and *Photoshop*.

In the summer before his senior year, Steve applied for and was offered a paid internship with a minor league baseball team in the Pittsburgh Pirates system. He said the electronic portfolio he'd been required to create in Desktop Publishing had helped him get the competitive position. He spent the summer working in the public relations department, doing everything from cleaning the stadium to redesigning the team website (and a revealing stint as a play-by-play announcer). At the end the internship he was offered a full-time position, which he had to decline in order to finish his senior year. But by the end of that fall, as he was completing his senior essay project, he'd accepted a full-time position as webmaster for another minor league team, to begin when he graduated.

Steve's senior essay project was tied directly to his internship experience and to his upcoming job, a visual and rhetorical analysis of every minor league baseball team website currently on line.

Two months into his employment, he emailed me asking for a reference for an application to a graduate program in sport management. I told him I was shocked—I'd thought he was living his career dream. He laughed and said he still was, he just discovered that he needed more training to move up. Steve would return to Ohio Northern to work part-time in the sports information office while working on his graduate degree. (A PW graduate from the previous year, a double major in PW and Sport Management, has also found career fulfillment in a university sports information office).

Ryan

Having graduated last year, a year later Ryan has yet to find a job. A sensitive, quiet, but funny and extremely likeable person, Ryan was an accomplished writer and reader. He completed a double major in PW and Sociology, but decided, halfway through his senior essay project, that he wanted to write fiction. He'd taken only one fiction writing course, an introduction to fiction writing, which counts as a PW elective. Ryan hadn't written any fiction since then. I was hesitant, but knew I had to approve the change in plans, pending department approval, even if it was a bit late in the game. After all, professional writers are supposed to be able to do it all. Ryan submitted a project proposal (a novella) and a writing sample to the department, which approved his request.

As Ryan's advisor, I agreed to work with him every week as he learned to write fiction. I was also honest with him about the slim career prospects for fiction writers. If he wanted to go on to graduate school in creative writing, he could, if he was lucky and he worked extremely hard, find a job teaching creative writing.

But had I, or the PW curriculum, misled Ryan, given him an inflated conception of his skills? Perhaps. But I was also wrong to think so. I've since reminded myself that we aren't a technical communication program. This isn't to absolve ourselves of the responsibility to prepare our students for viable careers; it is to say that our humanities touchstone remains central—a broad background rather than a narrow specialization. The truth is that a career as a fiction writer wasn't Ryan's goal anyway, even if fiction was the subject of his senior essay. Ryan's creative experience is as important and substantial in his professional training as a usability study.

Since beginning the PW program I'd carried around in my head Sullivan's and Porter's list of distinctions within the discipline of English—

Professional writing = writing for organizational forums; stress on corporate authorship
Creative writing = writing outside organizational forums (freelancing, whether "literary" or not); stress on individual authorship
Journalism = writing for public and mass-media forums
Writing in the academic disciplines = writing for disciplinary forums (i.e., to contribute to disciplinary knowledge) (412)

—and of my belief that my PW majors should have some experience with all of the above. Ryan may be a mediocre fiction writer, but from the correct perspective, he's an excellent professional writer. The correct perspective for an undergraduate PW program at a small university like Ohio Northern is double-sided: "Professional" indicates both a career orientation and a humanities-based generalism.

CONCLUSION

The unique qualities of a small university like Ohio Northern present unique advantages and disadvantages, limitations and opportunities, for an undergraduate professional writing program. Three main issues are of concern to us at this point in the progress of our program's five-year plan. In order of priority, they are:

Writing and Rhetorical Skills:
While the other four tracks (and every English department) have the same worry about student writing, our worry is particularly intense given the close relationship in the PW major of writing skills to professional opportunities. This is another point I emphasize to students in Writing in the Public Sphere: if you can't write well, you're not going to do well as a professional writer. Based on real-world demand for professional writing skills, Paul Bender and I want to emphasize in our writing courses audience awareness, knowledge of grammar, and critical thinking.

Real-World Connections:
Of all the PW program's connections to the workplace (publication practica, writing workshops, course service-learning projects), our internship requirement is the most important. Our students generally find good and useful internships, and we must continue to help students find them. We'll continue to expand our relationship with local, regional, and national internship opportunities.

Technology:
Paul Bender and I agree with Kim and Tolley, who have recently written about their own PTW program at the University of Memphis—technology is important for their students to learn, but only one aspect of their professional writing education (Kim and Tolley 385). Naomi has said that her company uses Adobe *InDesign* rather than *QuarkXPress,* and while she wishes she'd learned *InDesign* at ONU, she's learning what she needs to know on the job. Our graduates seem to be adequately prepared for workplace technology requirements, but it's up to us to keep pace with these requirements. It looks as if we'll be adding *InDesign* to our Desktop Publishing course at some point in 2006.

But while we must periodically upgrade our computer applications and our use of technology in our classrooms and courses, as a department we also agree that the emphasis of the PW program will continue to be on writing skills, rhetorical sophistication, and a critical cultural literacy. We'd like this foundation to mediate against the vagaries of real world technological, corporate, or market changes, even as we keep track of and respond to those—and a host of other—changes.

APPENDIX A

MAJOR IN ENGLISH/PROFESSIONAL WRITING [1998-2000]
(56 HOURS + SECONDARY STUDY)

QTR / YR √ COURSE COMPLETED   HOURS

Professional Writing Core   32
Required Courses

QTR / YR	√	Dept	Course	Hours
____/____	____	Art	222 Graphic Design 1	4
____/____	____	Comm	236 Public Relations Writing	4
____/____	____	Engl	243 Magazine Writing	4
____/____	____	Engl	251 Magazine Practicum	1
____/____	____	Engl	347 Advanced Writing	4
____/____	____	Engl	443 Nonfiction Writing	4
____/____	____	Engl	470 Editing	4
____/____	____	Engl	384 Directed Reading	1
____/____	____	Engl	483 Reading for the Senior Essay	1
____/____	____	Engl	484 Senior Essay 1	1
____/____	____	Engl	485 Senior Essay 2	2
____/____	____	Engl	481 Internship	1

Elective (Choose 1)   4

| ____/____ | ____ | Art | 223 Graphic Design 2 | 4 |
| ____/____ | ____ | Eng | 241 News Writing | 4 |

_____/_____ _____Comm 256 Telecommunications Writing
4
_____/_____ _____Engl 290 Special Topics (*in writing*)
4
_____/_____ _____Engl 342 Fiction Writing
4
_____/_____ _____Engl 343 Persuasive Writing
4
_____/_____ _____Engl 346 Prelaw Writing
4
_____/_____ _____Engl 377 Professional Writing Workshop
1-4
_____/_____ _____Engl 390 Special Topics (*in writing*)
4
_____/_____ _____Engl 451 Literary Criticism
4
_____/_____ _____Engl 490 Special Topics (*in writing*)
4

Language and Literature Core   20
_____/_____ _____Engl 210 English Studies
4
_____/_____ _____Engl 351 English Language
4

<u>Three literature courses in three core areas, two at the 300/400 level</u>
World Literature (Engl 208, 209, 219, 220, 262, 432, 451)   4
British Literature (Engl 213, 214, 260 *or* 412, 310-326, 364, 410)   4
American Literature (Engl 211, 212, 261, 334, 335, 365, 431)   4
*Other courses, whose content changes significantly with each offering, may
also satisfy these literature requirements: Engl 207, 263, 290, 390, 430, 490.*
_____/_____ _____
_____/_____ _____
_____/_____ _____

Engl 001 Enrichment (2 terms a year up to 8 times, depending on time enrolled as a major)
__/__ __/__ __/__ __/__ __/__ __/__ __/__ __/__

Secondary Study
An Option, Minor, or Second Major in a discipline other than English_____
                                                            (*more*)

APPENDIX B

MAJOR IN ENGLISH/PROFESSIONAL WRITING [2001-2004]
(57 HOURS + SECONDARY STUDY)

QTR / YR √ COURSE COMPLETED			HOURS
Professional Writing Core			**34**
Required Courses			30
_____/_____ 4	_____ Engl	244 Desktop Publishing	
_____/_____ 4	_____ Engl	243 Magazine Writing	
_____/_____ 4	_____ Engl	347 Advanced Writing	
_____/_____ 4	_____ Engl	405 Cultural Studies	
_____/_____ 4	_____ Engl	443 Nonfiction Writing	
_____/_____ 4	_____ Engl	470 Editing	
_____/_____ 1	_____ Engl	481 Internship	
_____/_____ 1	_____ Engl	384 Directed Reading	
_____/_____ 1	_____ Engl	483 Reading for the Senior Essay	
_____/_____ 1	_____ Engl	484 Senior Essay 1	
_____/_____ 2	_____ Engl	485 Senior Essay 2	

Elective (Choose 1)			4
_____/_____ 4	Engl	241 News Writing	
_____/_____ 4	Comm	256 Writing for Broadcasting & Electronic Media	
_____/_____ 4	_____ Engl	290 Special Topics (_in writing_)	
_____/_____ 4	_____ Engl	342 Fiction Writing	
_____/_____ 4	_____ Engl	343 Persuasive Writing	

| ____/____ | _____Engl | 346 Prelaw Writing |
4
| ____/____ | _____Engl | 377 Professional Writing Workshop |
1-4
| ____/____ | _____Engl | 390 Special Topics (*in writing*) |
4
| ____/____ | _____Engl | 451 Literary Criticism |
4
| ____/____ | _____Engl | 452 Rhetorical Theory |
4
| ____/____ | _____Engl | 490 Special Topics (in writing) |
4

Practicums (3 hours in at least two different practicums)   3
Engl 250-Newspaper; Engl 251-Magazine; Engl 230-Web Publishing; Engl 290-Journal Publishing; Engl 377-Professional Writing Workshop (1-4); Engl 2XX-Screenwriting

____/____	_____	1
____/____	_____	1
____/____	_____	1

Language and Literature Core   20
| ____/____ | _____Engl | 210 English Studies | 4 |
| ____/____ | _____Engl | 351 English Language | 4 |

<u>Three literature courses in three core areas, two at the 300/400 level</u>
World Literature (Engl 208, 209, 219, 220, 262, 432, 451)   4
British Literature (Engl 213, 214, 260 *or* 412, 310-326, 364, 410)   4
American Literature (Engl 211, 212, 261, 334, 335, 365, 431)   4
Other courses, whose content changes significantly with each offering, may also satisfy these literature requirements: Engl 207, 263, 290, 390, 430, 490.

____/____	_____
____/____	_____
____/____	_____

Engl 001 Enrichment (2 terms a year up to 8 times, depending on time enrolled as a major)
__/__ __/__ __/__ __/__ __/__ __/__ __/__ __/__

Secondary Study
An Option, Minor, or Second Major in a discipline other than English_____

(more)

APPENDIX C

MAJOR IN ENGLISH/PROFESSIONAL WRITING [2001–2006]
(58 HOURS + SECONDARY STUDY)

QTR / YR	√ COURSE COMPLETED			HOURS
Professional Writing Core				35
Required Courses				23
____/____	_____	Engl	244 Desktop Publishing	4
____/____	_____	Engl	273 Writing in the Public Sphere	4
____/____	_____	Engl	405 Cultural Studies	4
____/____	_____	Engl	445 Senior Seminar	2
____/____	_____	Engl	452 Rhetorical Theory	4
____/____	_____	Engl	470 Editing	4
____/____	_____	Engl	481 Internship	1
Electives (Choose three, two at the 300/400 level)				12
____/____	_____	Engl	241 News Writing	4
____/____	_____	Comm	256 Writing for Broadcasting and Electronic Media	4
____/____	_____	Engl	290 Special Topics (*in writing*)	4
____/____	_____	Engl	342 Fiction Writing	4
____/____	_____	Engl	343 Persuasive Writing	4
____/____	_____	Engl	344 Writing Cyberspace	4
____/____	_____	Engl	345 Screenwriting	4
____/____	_____	Engl	346 Prelaw Writing	4
____/____	_____	Engl	347 Advanced Writing	4
____/____	_____	Engl	390 Journal Publishing	4

_____/_____ 4	_____Engl	443 Nonfiction Writing
_____/_____ 4	_____Engl	447 Advanced Creative Writing
_____/_____ 4	_____Engl	451 Literary Criticism
_____/_____ 4	_____Engl	490 Special Topics (in writing)

Practicums (at least 3 hours in at least two different practicums) 3
Engl 250-Newspaper; Engl 251-Magazine; Engl 230-Web Publishing; Engl 231-Journal Publishing; Engl 377-Professional Writing Workshop (1-4); Engl 290-Screenwriting

_____/_____ _____ 1
_____/_____ _____ 1
_____/_____ _____ 1

Language and Literature Core 20

| _____/_____ | _____Engl | 210 English Studies | 4 |
| _____/_____ | _____Engl | 351 English Language | 4 |

<u>Three literature courses in three core areas, two at the 300/400 level</u>
World Literature (Engl 208, 209, 219, 220, 262, 432, 451)   4
British Literature (Engl 213, 214, 260 *or* 412, 310-326, 364, 410)   4
American Literature (Engl 211, 212, 261, 334, 335, 365, 431)   4
Other courses, whose content changes significantly with each offering, may also satisfy these literature requirements: Engl 207, 263, 290, 390, 430, 490.

_____/_____ _____
_____/_____ _____
_____/_____ _____

Engl 001 Enrichment (2 terms a year up to 8 times, depending on time enrolled as a major)

_/_ _/_ _/_ _/_ _/_ _/_ _/_ _/_

Secondary Study
An Option, Minor, or Second Major in a discipline other than English_____

MINORS, CERTIFICATES, ENGINEERING

8 Certificate Programs in Technical Writing: Through Sophistic Eyes

Jim Nugent

INTRODUCTION

Technical communication certificates are offered by many colleges and universities as an alternative to full undergraduate or graduate degrees in the field. Certificates typically require only one or two years of coursework strictly within technical communication, and typically can be earned while working full time or while seeking another degree. As Sherry Burgus Little notes in "Designing Certificate Programs in Technical Communication," certificate programs are diverse in their charter and construction. Some programs are geared toward those entering the field, while others are designed to augment the skills of practicing professionals. Some programs are designed to serve those in scientific and technical fields specifically, while others are designed to serve technical communicators more generally. Programs are offered at both the undergraduate and graduate levels, and the courses they require vary widely (Little 276–77).

According to the CPTSC/STC joint publication *Academic Programs in Technical Communication* (third edition), there were sixteen programs offering technical writing certificates in 1985 (Hayes 1). In 2003, the website of the Society for Technical Communication (STC) listed about eighty-four individual certificate programs, representing an approximate fivefold increase in number over eighteen years. Compared to the approximate doubling of the number of technical communication programs overall during the same period (Little 274), we can see certificate programs are an increasingly popular means of meeting the demand for skilled technical communicators.

Despite their increasing popularity, however, surprisingly little information or discussion exists in the current literature specifically about certificate programs. (Excepting, of course, the works presented for the first time in this volume, including Jude Edminster and Andrew Mara's valuable chapter "Reinventing Audience through Distance.") This informational and conversational void is unexpected, as certificate programs are currently situated in the middle of a number of related conversations in the field. With their vocational emphasis, certificate programs are potentially the site of conflict "on the issue of *training* opposed to *education,* or in other words, the conflict between theory and practice" (Little 278, emphasis in original). With their role in meeting the needs of local industry, and with their potential as the locations for academy-industry cooperation, certificate programs speak to the conversation of who shapes technical communication programs: academy or industry (Anderson, Bosley, Bushnell, Coon, Krestas, Yee, Zimmerman). With certificate programs' role as a gatekeeper to the profession—that is, the role of "certifying" implicit in their very name—they shape the professional identity of technical communicators and they present a number of significant implications to the project of professionalizing the field (Savage 364–5).

In this chapter, I seek to begin addressing the informational void surrounding certificate programs with the hope of making way for more productive dialog in the above conversations. In addition, I hope to provide some information, considerations, and cautions useful for administrators interested in implementing their own certificate programs. I begin the chapter by relating a two-part study of certificate programs that I performed in 2003. In the first part of this study, I examine sixty-two certificate programs to characterize them in terms of their curricular requirements. In the second part, I perform a survey of certificate program administrators to gauge who teaches in such programs, the age of such programs, and the relationship of such programs to local industry. Finally, I conclude by drawing from the work of Gerald Savage and others to suggest a potential framework for theorizing the certificate program in technical communication, namely sophistic rhetorical theory. Sophistic rhetoric, as Savage demonstrates, can act as a valuable tool for constructing the professional identity of the postmodern technical communicator. Likewise, I argue, it proves to be a valuable tool for theorizing the technical communication certificate program.

SITUATING MY APPROACH

Before I begin, I would like to step back for a moment and explain my (perhaps unusual) theoretical and methodological approach. In the course of completing and relating the work that follows, I have opted to employ a methodology itself informed by sophistic rhetorical theory. Such theory seeks to restore the reputation of the ancient sophists against the critiques of Plato. As part of a flourish of revived interest in rhetoric in general (see Jarratt, Poulakos, Crowley, McComiskey, and Leff), the revival of sophism embodies a potent critique of received Platonic and Aristotelian master narratives: narratives that inform the modernist belief in a coherent, external, and absolutely knowable reality set in opposition to a stable human subject.

The sophists were—and are—skeptics. Eschewing the foundationalist impulse toward fixed and transcendent Truths, the sophists maintain that a multiplicity of truths exist in any given context. The ancient sophists were travelers; they witnessed localized, multiple truths firsthand and they recognized how to work with them toward their own ends. However, despite their relativistic orientation, as Kenneth J. Lindblom notes, sophists avoid the paralyzing "trap of Pyhrronian skepticism"—"a nihilistic abyss of skepticism that refuses to regard even temporary truths" (93). The sophists are also teachers; the ancient sophists invented the profession of teaching and offered their instruction to any man who was able to pay their fees. Furthermore, the sophist is socially engaged. A sophist believes that the key to meaning-making lies in our social interactions, and not in an abstract realm of Platonic ideals (Jarratt, Leff, Lindblom). The sophists, then, provide a very productive metaphorical and historiographic embodiment of a broader postmodern critique.

Drawing from Susan Jarratt's work on sophistic historiography ("Toward"), Bruce McComiskey suggests that "A certain view of historiography goes along (or *should* go along) with the neosophistic goal of appropriation and methods of mining and transporting doctrines, a view [itself] based on sophistic principles" (56, emphasis in original). As a researcher, then, I have attempted to remain aware that the historical instance that I am examining "does not exist in any *essential* form," and that even if it did, I "can not know it except through the process of interpretation" (McComiskey 56, emphasis in original). My assertions in this study, therefore, do not "strive for cognitive certitude, the affirmation of logic, or the articulation of universals" (Poulakos 37), but rather are grounded in conversations within the field of technical communication and in the needs of those working within the discipline. Put simply, in this chapter I make no pretense of offering anything other than partial, contingent, assailable, contextual, localized—and hopefully useful—knowledge.

Throughout this chapter, I attempt to avoid what Donna Haraway terms the "god-trick": the act of producing knowledge that pretends "to be from everywhere and so nowhere, to be free from interpretation, from being represented, to be fully self-contained or fully formalizable" (196). Although I use decidedly empirical methodologies, I make some departures from their traditional implementation. In Part I, for instance, I develop a heuristic for categorizing the courses in the certificate programs I look at. However, in no way do I conceive of this heuristic as taxonomic or representative: it is not a bijective mapping between the spaces of objective reality and a higher Platonic realm. Rather, I recognize that this heuristic is a product of my perspective on certificate programs, and it simultaneously—not to mention paradoxically and recursively—shapes that very same perspective.

I have done my best to avoid the rhetorical moves common to "god-trick" scientific narratives: I try to resist using the passive voice to disguise my own interpretive role; I attempt to avoid the familiar conceits of the scientific report genre which serve to obscure its own constructedness; and I attempt to bracket my findings not as articulated universals, but as paths for moving forward in conversation. These ethical moves are what I believe to be necessary to remain consistent with the situated and contingent practice of technical communication as a sophistic profession.

PART I: LOOKING AT CERTIFICATE PROGRAM CURRICULA

As I mentioned earlier, little information exists in the literature about technical communication certificate programs (see Bosley, Hayes, Bridgeford, Little). Little's chapter in the 1997 volume *Foundations for Teaching Technical Communication* is perhaps the most expansive work specifically on certificate programs. Here, Little provides a general review and synthesis of the information on certificate programs found in the four editions of *Academic Programs in Technical Communication* published between 1976 and 1993. While this overview provides useful conclusions about the diversity of certificate programs, it does not attempt a systematic or detailed characterization of them. In this part of the study, I attempt to gauge in greater detail what certificate programs in technical communication require in terms of coursework.

The Society for Technical Communication (STC) maintains an online Academic Programs Database containing information on technical communication programs at all levels, and which served as the origin of this study. This database is publicly available at http://www.stc.org/academicDatabase.asp, and the data held there include the name of programs, the department that houses them, contact information (address, phone, email, and homepage URL), the programs and degrees offered, the number of credits required to graduate, the approximate time of completion, the number of students in the program, the number of graduates per year, and a brief description of the program. In the spring of 2003, I used the database's web interface to select all programs listed as offering a certificate program, yielding 122 records. I copied the data from those records into a Microsoft Access database.

Next, I attempted to visit the websites for all 122 programs using the URLs provided in the database, or by performing a general web search when

I found no working URL. My initial goal was to determine which programs I would include in the study, and then later to collect program curricula and course descriptions from online sources. I selected for inclusion programs that met all of the following criteria:

- The certificate program was expressly in technical communication or technical writing
- The certificate was an independent degree, and was not required to be earned concurrently with another degree as a minor would be (however, programs could prerequire a degree for admission to the program)
- Sufficient information was available online to determine the program's curriculum and course requirements
- The program information was available in English and courses were taught in English

Of the 122 initial records, I determined that sixty-two met the above criteria for inclusion. Of the sixty records that I excluded:

- Six were duplicates of other records
- Thirty-two were misidentified as certificate programs or offered no identifiable certificate program in technical communication
- Nine were for programs that required a concurrent degree (specifically, a bachelor's degree)
- Nine were for programs that did not have sufficient program information available online to determine their curricula
- Four were for programs in a foreign language

I should note that these criteria may be the source of possible selection bias in the results that follow. By excluding nine programs for not having sufficient information online, for instance, I may have encouraged an overrepresentation of digital technology courses, as the lack of sufficient web presence at those institutions may conceivably also reflect the lack of major technology initiatives, training, or funding. In addition, by limiting the study to independent certificate programs (which do not require a concurrent degree), I may have also encouraged a small overrepresentation of industry-connected programs in the surveys for Part II, below. Such programs, lacking the "captive audience" of an undergraduate student body, may have a greater incentive to recruit students and feedback from local industry. Although I don't feel that these decisions significantly impair the usefulness of the data I developed, I do feel that a sophistic approach compels me to point out these possible biases here.

For each of the sixty-two certificate programs I identified, I obtained the requirements for completion from the institution's website or online course catalog. Next, I compiled a list of every one of the 863 non-unique course titles counting toward certificates in these programs. By iterating through this list several times, I developed a course title coding heuristic that identified nine major categories of courses: General Technical Communication/Technical Writing, Technical Communication Genre Writing, Other Writing, Editing, Communication and Rhetoric, Management, Visual Design, Digital Technology, and Miscellaneous. From these categories, I further identified subcategories, and where necessary, sub-subcategories. By making the heuristic hierarchical in this way, I was able to adapt the level of particularity to meaningfully group courses together, while still being able make claims about broader trends across programs. For instance, the Title Software subcategory of Digital Technology is for courses dedicated solely to specific titles of software. In order to determine further what kind of software is being taught—while still maintaining Title Software as a meaningful course category—I created sub-subcategories for the specific types of software titles: Layout and Publishing, Digital Graphics, Documentation and Help, etc.

After refining the heuristic to contain over sixty categories and subcategories—an early indication of the breadth of course offerings in certificate programs—I went on to examine the individual curricula for each of the sixty-two programs, coding their course titles according to the heuristic. In addition to coding the category courses belonged to, I also noted whether they were required or elective according to the following definitions:

- Required courses are courses that are required of all certificate students and are not elective
- Elective courses are courses that are selectable from a list of two or more

In Table 1, I show the data I developed in the course of this study. In the left column, I list the categories and subcategories of the course heuristic. The first column (#R) lists the number of surveyed programs I found requiring at least one course in that category, and the second column (%R) lists the same quantity expressed as a percentage of the sixty-two surveyed programs. Similarly, the third column (#E) lists the number of surveyed programs I found offering at least one course in that category as an elective and the fourth column (%E) lists the same quantity expressed as a percentage of the sixty-two surveyed programs.

The certificate programs I examined require on average 7.6 courses for completion, but they range in number from three to fourteen courses. While a vast majority of the programs follow a typical fifteen-week university semester schedule, several follow a different schedule according to the policy of their institution—a small few of which are non-academic. Since my primary focus was on the topic of instruction in certificate programs, I made no attempt to record the length of time required for program completion.

The first and most immediately apparent finding of this study, I believe, is that certificate programs include a very wide variety of courses in an equally wide variety of curricula—confirming Little's 1997 conclusions regarding program diversity (276). This is supported by the fact that, in order to meaningfully categorize program courses, I had to develop a heuristic with over sixty different course types. Of these course types, I was unable to identify a single one common to each and every certificate program curriculum, whether as a requirement or as an elective. In addition, I found only one broad course type that is required by a majority of certificates. With such a disparate makeup of programs, I maintain, there is no such thing as a "standard" or "core" technical communication certificate curriculum.[1]

But despite the breadth of certificate programs' course offerings, I found some courses to be clearly more popular than others. The most popularly required courses, I argue, could be said to represent a "not uncommon cluster." These courses fall into the following categories:

1. General technical communication (including introductory and advanced technical communication/writing service courses)
2. Editing
3. Technical communication genre courses
4. Digital technology

The most commonly required courses are in general technical communication, with 71% of surveyed programs (44) requiring at least one. The advanced general course—which frequently goes by the title Technical Communication II or Advanced Technical Writing—is the only specific course subcategory that I found to be required in a majority of the programs surveyed, with 56% of programs (35) requiring it. The next most commonly required courses are in editing, with 45% of programs (28) requiring at least one course in some form of editing (including the subcategories of technical editing, copyediting/proofreading, and grammar). After editing, the most commonly required group of courses are those dedicated solely to a specific genre of technical writing such as reports, procedures, and proposals, with 40% of identified programs (25) requiring at least one course. The most popular genre courses are dedicated to manuals and procedures (with 18% of programs requiring at least one) and computer documentation (with 16% of programs requiring at least one). Finally, over a third of programs surveyed (22) require at least one course related to digital technology, with courses dedicated to specific titles of software making up the largest portion of both required and elective offerings.

The courses I classified into the "Miscellaneous" category are those that were difficult to meaningfully include in other groups. The most prevalent such courses include projects and practicum courses (with 15% of programs requiring at least one course), internship courses (13%), and courses in usability or human factors (13%). Also within the "Miscellaneous" group, I found that some of the least commonly required certificate program courses provide an interesting glimpse at the competing priorities for certificate programs. A very small minority of the surveyed programs' curricula reflect the historically common situation of technical communication within departments of English: 3% of programs (2) offer a course in literature as an elective, 6% of programs (4) offer a course in creative writing as an elective, and 3% of programs (2) offer a course in the teaching of writing as an elective. While none of these programs went so far as to explicitly require these offerings, they nonetheless reflect the priorities of an English studies curriculum and not necessarily those of technical communication.

However, I found that the least commonly required courses also suggest a different, emerging set of priorities for technical communication certificate programs. One program requires a course in the history of technical communication for certificate completion, and two programs offer such a course as an elective. Meanwhile, two programs require a course in law or ethics, and 8% of programs (5) offer a similar course as an elective. The very least common courses—those that were so singular that they defied classification as anything besides "Miscellaneous–Other"—included course titles such as: Job Search for Technical Writers, Applied Psychology of Technical Communication, Starting a Technical-Writing Career, People Skills for Technical Writers, and Marketing Technical Writing Services. These courses reflect a surprising specificity to individual certificate program curricula, as well as reflect the diversity of technical communication as an emerging discipline.

In summary, I found that the technical communication certificate programs I surveyed vary widely, and no core curriculum can be said to exist among them. The programs I looked at in this study, although disparate and wide-ranging, clearly favor four major groups of courses, which we can consider to be a "not uncommon cluster" of certificate program courses: general courses in technical communication (including the technical communication/writing service courses, particularly the advanced course), courses in editing, courses devoted solely to specific technical communication genres, and courses in digital technology. Finally, while the least commonly required courses can't be said to reflect the pedagogical priorities of certificate programs most broadly, I believe that they may reflect those of either the past (those of an English studies curriculum) or the future (those of an emerging technical communication discipline).

COURSE CATEGORY	#R[1]	%R[2]	#E[3]	%E[4]
Technical Communication/Technical Writing	44	71%	26	42%
Introductory General Course	16	26%	3	5%
Advanced General Course	35	56%	10	16%
Other General Course	17	27%	16	26%
Technical Communication Genre Writing	24	39%	25	40%
Reports	6	10%	8	13%
Manuals and Procedures	11	18%	7	11%
Computer, Software, and Online Documentation	10	16%	22	35%
Proposals	6	10%	10	16%
Portfolio and Résumé	1	2%	1	2%
Other Writing	18	29%	22	35%
Composition	12	19%	5	8%
Introductory Service Course	10	16%	2	3%
Advanced Service Course	3	5%	1	2%
Other	1	2%	4	6%
Business and Professional Writing	9	15%	7	11%
Science and Medical Writing	1	2%	7	11%
Journalism and Newsletter Writing	2	3%	9	15%
Creative Writing	0	0%	4	6%
Other	2	3%	9	15%
Editing	28	45%	27	44%
Technical Editing	15	24%	8	13%
Copy Editing and Proofreading	10	16%	6	10%
Grammar	8	13%	6	10%
Other	0	0%	1	2%
Communication and Rhetoric	17	27%	33	37%
Speech and Presentation	6	10%	12	19%
Business and Professional Communication	2	3%	10	16%
Public Relations	1	2%	5	8%
Marketing and Advertising	0	0%	6	10%
Interviewing	2	3%	2	3%
Training and Instruction	2	3%	6	10%
Rhetoric	4	6%	4	6%
Other	2	3%	8	13%

(more)

Management	17	27%	8	13%
Project Management	11	18%	5	8%
Organizational Management	2	3%	3	5%
Information and Knowledge Management	6	10%	3	5%
Other	0	0%	2	3%
Visual Design	19	31%	20	32%
Layout	9	15%	9	15%
Graphics and Graphic Design	6	10%	10	16%
Multimedia	2	3%	3	5%
Photography	0	0%	2	3%
Illustration	0	0%	1	2%
Other	2	3%	5	8%
Digital Technology	22	35%	20	32%
Introductory	6	10%	4	6%
General Web	4	6%	10	16%
Desktop Publishing	6	10%	8	13%
Title Software	11	18%	14	23%
Layout and Publishing	0	0%	2	3%
Digital Graphics	0	0%	5	8%
Documentation and Help	5	8%	8	13%
Word Processing	3	5%	2	3%
Web	3	5%	3	5%
Other	2	3%	0	0%
Programming Languages and Protocols	1	2%	6	10%
Database and Information Technologies	2	3%	4	6%
Other	4	6%	7	11%
Miscellaneous	21	34%	27	44%
Usability and Human Factors	8	13%	5	8%
History of Technical Communication	1	2%	2	3%
Internship	8	13%	10	16%
Projects and Practicums	9	15%	6	10%
Law and Ethics	2	3%	5	8%
Linguistics	0	0%	2	3%
Literature	0	0%	2	3%
Mathematics	1	2%	0	0%
Print Production	1	2%	5	8%
Research and Critical Thinking	0	0%	9	15%
Teaching Writing	0	0%	2	3%
Other	5	8%	12	19%

TABLE I: THE DATA I DEVELOPED IN PART I

[1] The number of surveyed programs requiring at least one course
[2] #R expressed as an approximate percentage of the sixty-two surveyed programs.

[3] The number of surveyed programs offering at least one course as an elective.
[4] #E expressed as an approximate percentage of the sixty-two surveyed programs.

PART II: SURVEY OF PROGRAM ADMINISTRATORS

During the fifteenth annual meeting of the Council for Programs in Technical and Scientific Communication (CPTSC) in 1988, the workshop group on certificate programs recommended that the council perform a nationwide survey to "gather information on the context of existing Certificate programs" and to determine the status of instructors in such programs (Hayes 29). In Spring 2003, I sought to respond to this long-unheeded call by surveying the administrators of certificate programs included in Part I of the study. My survey included questions on the size of the program; the status, qualifications, and specialization of instructors in the program; the age of the program, and the relationship of the program to local industry. The survey was sent by email to the contacts specified in the STC database, and of the sixty-two surveys I sent, 42% (twenty-six) were returned complete. The results are summarized as follows:

- Most certificate program instructors are required to have at least a master's degree, and much less frequently, a doctorate. A majority are required to have experience in industry as well. When asked what qualifications were required of their certificate program instructors,
    - 96% of respondents (25) indicated a bachelors degree
    - 85% of respondents (22) indicated a master's degree
    - 31% of respondents (8) indicated a PhD
    - 62% of respondents (16) indicated industry experience
    - 4% of respondents (1) indicated other experience, specifically, "Experience with relevant software or markups such as FrameMaker, RoboHelp, HTML, Word, Powerpoint, Acrobat, Dreamweaver, Photoshop depending on course."
- The mean reported age of certificate programs (in 2003) was 10.4 years, with programs ranging in age from one to twenty-two years
- 54% of respondents (14) indicated that their program makes use of an industry advisory board, while 42% (11) indicated that they do not
- 58% of respondents (15) indicated that their program actively recruits from local industry
- 62% of respondents (16) indicated that their program employs other procedures or mechanisms to gather feedback from local industry, while 35% (9) indicated that they do not

- 38% of respondents (10) indicated that their program requires work in industry as a part of courses required for program completion
- When asked how industry feedback is solicited, the most commonly specified sources were: professional associations such as the STC or the CPTSC (five respondents), followed by feedback from students (four respondents), feedback from internship partners (four respondents), guest lecturers (four respondents), program-sponsored events (three respondents), and alumni contacts (two respondents).

In summary, I found that almost all of the responding certificate programs require a master's degree or better for their program instructors, and a strong majority require instructors to have had industry experience. Most significantly, I found that the programs I surveyed demonstrate close ties to industry: most programs actively recruit from local industry, maintain industry advisory boards, or employ other less formal mechanisms for industry feedback. However, only 38% of programs (10) indicated that they require students to actually work in industry for program completion.

PART III: SOME POSSIBLE IMPLICATIONS

The data and findings I develop here only begin to address the informational void surrounding technical communication certificate programs. In an attempt to draw this work closer into existing conversations, I would like to conclude this chapter by suggesting a potential framework for theorizing the certificate program. As Gerald J. Savage demonstrates in his chapter "Tricksters, Fools, and Sophists: Technical Communication as Postmodern Rhetoric," the sophist provides a compelling model for the identity of the technical communicator:

> [T]he work of technical writing seems to be consistent with a sophistic practice in which knowledge is always contingent, in which rhetorical purpose must be reconciled to the needs of a particular audience at a particular time and place. Technical writing as we find it today has emerged in relation to particular economic, political, and technological circumstances which combine in complex and contradictory ways that make the work our practitioners do both useful and disruptive, both materially rewarding and risky [...] Yet these circumstances present us with the strongest argument for accepting the apparently weak role of the non-expert, unrecognized, incompletely professionalized, uncertified, hard to define sophist-technical communicator. (189)

By situating technical communication as a sophistic profession, its members become "politically and socially engaged communicators who recognize the inevitability of their texts as socially transformative" (171). Its members also embrace their status as "liminal subjects," "occupying marginal zones between the subject matter expert and the lay audience, trading status both in the corporation and in larger society for relative freedom to travel across the boundaries of these social-cultural domains" (180).

Certificate programs in technical communication can be seen as occupying a similar liminal zone: that between academy and industry; theory and practice; education and training; local and universal. Just as sophism concentrates on the individual locations and contexts of knowledge production at the expense of universal precepts and monolithic Truths, certificate programs often situate themselves to meet the practical needs of local industry at the expense of teaching more generalizable academic theory (Little 278). Just as sophism threatens the modernist distinction between theory and practice (Leff 24, Scott 193), the situation of the certificate program between academy and industry lends it the potential to collapse theory and practice into postmodern *praxis*. Certificate programs, I assert, are theoretically consistent with—and are poised to support the work of—technical communication as a sophistic profession.

Existing programs, as I explored in this chapter, support this assertion. In Savage's characterization of technical communication as sophistic, the field avoids the modernist impulse to gain professional status and market closure. At the expense of achieving fixity in its professional identity and knowledges, then, the field gains the ability to remain flexible in the face of an ever-changing postindustrial workplace (188–9). As I show in this chapter, existing certificate programs display a programmatic and curricular flexibility that is consistent with this characterization. In Part I, I suggest that the curricula of certificate programs are so wildly disparate that no core curriculum can be said to exist among them. And in Part II, I show that the certificate programs I surveyed display strong ties to local industry, with most employing some means of soliciting industry feedback. These findings together are consistent with a conception of technical communication as a postmodern profession in a market where no standard, universally-required skill set has emerged—whether from the collective needs of industry or as a result of professionalizing gestures from elsewhere.

However, while technical communication certificate programs are decidedly sophistic in their flexibility, this very feature makes them fraught with the potential for appropriation. By embracing flexibility alone, certificates could easily become "value added" degrees that serve the most immediate material interests of schools and students while failing to provide any relevant preparation for the situated practice of technical communication. Fortunately, sophistic theory addresses this concern by suggesting a number of vital caveats for certificate program design. Platonic curricula—which sophistic curricula can be said to be articulated against—would hold that the practices of technical communication are entirely reducible to formalizable first principles, and therefore such curricula only demand sufficient classroom time to facilitate the "transfer" of formalized knowledge from teacher to student. By contrast, a thoroughly sophistic curriculum must recognize that the practice of technical communication is contingent, localized, and social, and should therefore make space beyond the classroom for students to develop appropriate professional capacities in context. In other words, a sophistic curriculum demands social engagement.

As Susan Jarratt notes in *Rereading the Sophists,* "the sophists could be termed the first public intellectuals in a democracy" (98). Sophism is, by its nature, publicly accountable and "immersed in the adjudication of immediate cultural concerns" (Crowley 318), an attribute that Savage foregrounds in his own characterization of the sophist technical communicator. An important consequence of this social orientation, I believe, is that sophistically conceived certificate programs must include opportunities for students to take their work beyond the walls of the academy. Although I found that the programs in this study demonstrate a commitment to the interests of local industry, the fact that only 38% (10) of the surveyed programs in Part II require students to work in industry for program completion suggests that, at least at a curricular level, certificate programs could do more to prepare students for their social roles in a sophistic profession. This thesis is further corroborated by the findings in Part I: only 13% (8) of the sixty-two certificate programs I surveyed require an internship for course credit, and only 18% (9) require a project or practicum for course credit. Nonetheless, it remains to be seen if programs enact social engagement at other levels: for instance if students already work extensively in local industry, or if individual courses and pedagogical methods already emerge from local needs.

Another strategy that sophism suggests for program design is the incorporation of reflexive professional development. When seen as a sophistic profession, the qualification of a technical communicator is not a discrete skill set that he or she possesses; rather it is a professional *ethos* that he or she has developed. Phrased another way, the identity of the sophist–technical communicator can be seen not so much as a subjectivity (one who possesses knowledge in the Platonic sense), but rather as an *intersubjectivity* (one possessed of a certain *ethos,* or way of acting within and among social realms). This intersubjectivity is not assumable by rote and it cannot be taught through a Platonic curriculum of disjointed courses; it must be developed instead by allowing students to make the connections between their coursework and the social realm of technical communication in practice. This development can be facilitated, I believe, through self-conscious reflexivity.

Such reflexivity is comprised of an explicit engagement on behalf of the student with the issue of what it means to be a practitioner of technical communication. It can take the curricular form of retrospective portfolios, capstone projects, student symposia, close instructor advising, or even developmental journals; but the end goal of these activities should be for students to self-consciously adopt the professional *ethos* of a technical communicator within—and as shaped by—their specific social and cultural contexts. To be sure, such reflexive practice is conceived here not as a Platonic act of philosophical contemplation but as a sophistic act of rhetorical engagement: each of these activities should be constructed with a genuine audience, purpose, and context. Through these kinds of reflexive activities, students are provided with the curricular space to make developmental connections through social and discursive means. Although it is not certain the extent to which the kind of reflective professional development that I am sketching here is a part of current certificate programs, it remains an intriguing issue for future research, as well as a compelling consideration for the design of any technical communication curriculum.

I hope I have made clear that I do not see the value of sophism as a means to excuse programmatic configurations that are convenient, but otherwise inexcusable. Rather, in offering a model of professional identity as an alternative to those of market closure and fixity, sophistic rhetoric provides a more situated and responsible figuration for the technical communicator. As I find in this chapter, certificate programs in technical communication show great potential as the sites for realizing this sophistic model. My research on existing certificates shows that they are consistent with a sophistic model of programmatic flexibility and concern for local needs. However, the sophistic model also provides important cautions for the design and administration of certificate programs. A sophistically-informed certificate program must remain socially engaged, and it should present opportunities for students to work in real social contexts. In addition, a sophistically informed certificate program must be reflective: it should present opportunities for students to reflexively and self-consciously develop their professional *ethos*.

Again, I make no claims to transcendence or universality in the research and interpretations I present here, and I hope that I have avoided any of the familiar rhetorical techniques that would suggest otherwise. I also hope that the information and discussion I have presented are useful—either for participants in conversations within the field or for those considering the development of their own certificate programs. Sophistic rhetorical theory, I believe, provides an invaluable theoretical model—one that is both descriptive and prescriptive—for building and understanding certificate programs in technical communication.

ACKNOWLEDGMENTS

Of course, my sophistic approach compels me to recognize the social origins of this work. Thanks to Gerald Savage for his invaluable guidance and help with the program research; thanks to Ken Lindblom for his assistance; and thanks to Ann Brady, Robert Johnson, Lori Ostergaard, Anne Parker, and Marika Seigel for their responses to drafts.

NOTES

[1] An identical conclusion was reached about baccalaureate degree programs in a comparable 2005 study by Sandi Harner and Anne Rich. There they conclude: "It is clear from this study that there is no standard curriculum for technical communication programs" (219).

WORKS CITED

Anderson, Paul V. "Evaluating Academic Technical Communication Programs: New Stakeholders, Diverse Goals." *Technical Communication* 42.4 (1995): 628–633.

Bosley, Deborah S. "Broadening the Base of a Technical Communication Program: An Industrial/Academic Alliance." *Technical Communication Quarterly* 1.1 (1992): 41–50.

Bosley, Deborah S. "Building a Graduate Certificate Program: Making Connections With the Community-at Large." *Proceedings of the Annual Meeting of the Council for Programs in Technical and Scientific Communication (24th, Austin, Texas, October 16–18, 1997)*. Socorro, NM: New Mexico Institute of Mining and Technology, 1998.

Bosley, Deborah S. "Collaborative Partnerships: Academia and Industry Working Together." *Technical Communication* 42.4 (1995): 611–619.

Bridgeford, Tracy. "Thoughts on Designing a Master's Certificate Program." *Models for Strategic Program Development. Proceedings of the Annual Meeting of the Council for Programs in Technical and Scientific Communication (27th)*. Ed. Bruce Maylath. Menomonie, Wisconsin, 19–21 Oct, 2000. http:// www.cptsc.

org/conferences/conference2000/program2000/issuesfacingmastersprograms/bridgeford.pdf.

Bushnell, Jack. "A Contrary View of the Technical Writing Classroom: Notes Toward Future Discussion." *Technical Communication Quarterly* 8.2 (1999): 175–188.

Coon, Anne C., and Patrick M. Scanlon. "Does the Curriculum Fit the Career? Some Conclusions from a Survey of Graduates of a Degree Program in Professional and Technical Communication." *Journal of Technical Writing and Communication* 27.4 (1997): 391–399.

Crowley, Sharon. "A Plea for the Revival of Sophistry." *Rhetoric Review* 7 (1989): 318–334.

Haraway, Donna J. *Simians, Cyborgs, and Women: The Reinvention of Nature.* New York: Routledge, 1991.

Harner, Sandi, and Anne Rich. "Trends in Undergraduate Curriculum in Scientific and Technical Communication Programs." *Technical Communication* 52.2 (2005): 209–220.

Jarratt, Susan C. *Rereading the Sophists: Classical Rhetoric Refigured.* Carbondale: Southern Illinois UP, 1991.

Jarratt, Susan C. "Toward a Sophistic Historiography." *PRE/TEXT: The First Decade.* Pittsburgh Series in Composition, Literacy, and Culture. Ed. Victor J. Vitanza. Pittsburgh: U of Pittsburgh P, 1993. 263–85.

Krestas, Shirley A., Lori H. Fisher, and JoAnn T. Hackos. "Future Directions for Continuing Education in Technical Communication." *Technical Communication* 42.4 (1995): 642–645.

Leff, Michael C. "Modern Sophistic and the Unity of Rhetoric." *The Rhetoric of the Human Sciences.* Eds. Allan Megill, John S. Nelson, and Donald N. McCloskey. Madison: University of Wisconsin Press, 1987. 19–37.

Lindblom, Kenneth J. "Toward a Neo-Sophistic Writing Pedagogy." *Rhetoric Review* 15.1 (1996): 93–107.

Little, Sherry Burgus. "Designing Certificate Programs in Technical Communication." *Foundations for Teaching Technical Communication: Theory, Practice, and Program Design.* Ed. Katherine Staples and Cezar Ornatowski. Greenwich, CT: Ablex, 1997. 273–85.

McComiskey, Bruce. "Neosophistic Rhetorical Theory." *Gorgias and the New Sophistic Rhetoric.* Carbondale: Southern Illinois UP, 2002. 55–76.

Poulakos, John. "Toward a Sophistic Definition of Rhetoric." *Philosophy and Rhetoric* 16 (1983): 35–48.

Proceedings of the Annual Meeting of the Council for Programs in Technical and Scientific Communication (15th, Minneapolis, Minnesota, October 19-21, 1988). Ed. Laurie Schultz. Minnesota, 1988.

Savage, Gerald J. "The Process and Prospects for Professionalizing Technical Communication." *Journal of Technical Writing and Communication.* 29.4 (1999): 355–381.

Savage, Gerald J. "Tricksters, Fools, and Sophists: Technical Communication as Postmodern Rhetoric." *Power and Legitimacy in Technical Communication, Volume II: Strategies for Professional Status.* Ed. Teresa Kynell-Hunt, and Gerald J. Savage. Baywood's Technical Communications Series. Amityville, NY: Baywood, 2003. 167–193.

Scott, Blake J. "Sophistic Ethics in the Technical Communication Classroom: Teaching *Nomos,* Deliberation, and Action." *Technical Communication Quarterly* 4.2 (1995): 187–199.

Society for Technical Communication Academic Programs Database. 2004. Society for Technical Communication. August 2004 <http://www.stc.org/academic-Database.asp>.

Yee, Carole. "Can We Be Partners?: Industry/Corporate Advisory Boards for Academic Technical Communication Programs." *Publications Management: Essays for Professional Communicators.* Ed. Jane O. Allen and Lynn H. Deming. Baywood's Technical Communications Series. Amityville, NY: Baywood, 1994.

Zimmerman, Donald E., and Marilee Long. "Exploring the Technical Communicator's Roles: Implications for Program Design." *Technical Communication Quarterly* 2.3 (1993): 301–317.

9 Shippensburg University's Technical / Professional Communications Minor: A Multidisciplinary Approach

Carla Kungl
S. Dev Hathaway

To meet increasing student interest in technical and professional opportunities for post-graduate career preparation, Shippensburg University, PA, has recently developed an interdisciplinary Technical/Professional Communications Minor, bringing together a variety of pertinent courses from across the college curriculum and organizing them into a minor that is substantive, coherent, and flexible. This article discusses the development and implementation of Shippensburg University's new interdisciplinary minor. While the program's final structure sprung primarily from necessity, its multidisciplinary status will allow our students to reap many unforeseen benefits. We feel this program could be a successful model for other smaller schools to follow, schools that have neither the student numbers nor the resources to begin their own majors or minors in technical communications.

BACKGROUND: WHO WE ARE

Shippensburg University is a public university in the State System of Pennsylvania of about 6600 students. We attract an increasingly competitive student body, with SATs averaging about 1100. Most are from across Pennsylvania, and a number are from rural areas and are first-generation college students, who especially see college as an important and vital step in preparing for a professional career. At the same time, our university is committed to its traditional liberal arts curricula: we maintain a strong general education program, expose students to many fields of inquiry, and encourage close faculty-student relationships and community service.

Thus, though the English Department felt pressures (from an array of sources which I'll discuss) to provide a more "professional" education—to teach more overtly the workplace skills our students might need—our faculty believed whole-heartedly that our liberal education was best suited to instill the knowledge, judgment, and skills base for all college-educated citizens, for the greater community as well as the professional workplace.

In determining how to supply our graduates with the higher-education communications skills we felt they needed to succeed in professional careers, we looked to contributions from across the university and developed a multidisciplinary model to help give students access to a range of courses and skills. This model allows us to marshal resources and course offerings heretofore segregated among various college departments. In so doing, it provides a career-enhancing program for students while maintaining a meaningful liberal arts backdrop.

STRIKING A BALANCE: THE PRESSURE TO PROFESSIONALIZE VS. MAINTAINING A LIBERAL EDUCATION

One of the most obvious pressures to update our students' skills came from outside academia—the need to meet demands of the information technology job market. Professionals capable of combining technical expertise with communication skills are sought after, and the increased numbers for and roles of technical communicators in various industries has therefore impacted university instruction. Aimee Whiteside, in her survey of the skills that the new generation of technical communicators need, reiterates the feeling that the working world's rapid changes "created a profound challenge for academia, which grappled to balance pedagogical strategies and foundational critical thinking skills with specific skills that technical communication students need to be successful in business and industry" (303). Universities have tried different strategies for giving their graduates the necessary skills: some have added whole departments in technical or professional writing; some have implemented new majors and minors in technical writing or communication; some have created writing tracks within an existing department (in our own geographical area, for example, the number of major or minor programs has nearly doubled in the past five years).

But for many schools it is hard to know exactly what kind of program or how much of a program to create to fulfill the demands of their community. They then run the risk of leaving either professional or pedagogical gaps in their students' education, or both. Rude and Cargile Cook's recent article on the academic job market in technical communication discusses this problem in light of the issue of an inadequate number of trained faculty to teach this burgeoning population of students. But the assumptions governing the fates of both groups, faculty and students, stem from the same set of problems: uncertain job markets and difficulties adequately assessing future need "on the basis of current demand" (50). Their point that "the growth of academic programs [in PTW] and the parallel demand for new faculty seem tied to growth of the role for technical communicators in the corporation" is a nice piece of information, but one that is hard to address logistically (50). What's a small regional school to do?

The other major pressure we were feeling came from the State System Chancellor's Office. Like most program innovations, our Technical/Professional Communications Minor arose several years ago from discussions effected by mission and curricular changes already in the air if not in the works. Our administration was increasingly underscoring the need for competitive professional preparation programs (though offering little in the way of additional resources to underwrite them). In response to this, for example, the History Department retooled their Master's Degree to offer an MA in "Applied History," providing training in more practical applications of a History degree: how to be an archivist, or a tour guide, or a curator. Our own MA came under the hatchet at about the same time because we couldn't come up with a suitable way to make it more professionally oriented (now we offer just a few graduate courses per year in the Department of Education's Curriculum and Instruction degree).

We also felt pressures a little closer to home. In the fall of 1999 our English Department underwent a constructive five-year review. One of the suggestions that came out of that review was to revamp our degree programs to include, among other new features, a Writing Emphasis option to accommodate those students who wanted more practical skills but who were not interested in teaching (nearly half of our majors are Secondary Education students). Our new writing track came with a commitment to hire faculty with expertise in technical writing and to develop two new courses, Technical/Professional Writing I & II. By creating this Writing Emphasis, we saw a way not only to supply students with new courses but also to provide a new professional slant on existing courses, such as Reviewing the Arts.

While serving as department chair at that time, Prof. Hathaway had concerns of two opposing kinds. First, in agreement with the department's outside reviewer, he saw the need to better prepare the majority of our English graduates for their immediate future in the professional workplace. But at the same time, he and the rest of the department wanted to stem the "workforce prep" mentality that we felt was beginning to threaten the liberal arts heart of higher education in the State System. He discussed options with the now-former Dean of College of Arts and Sciences, who, though appreciative of our dilemma, communicated the State System's increased professional preparation emphasis. At the same time, she was very supportive of realigning our options in the major and adding Technical Writing courses; she also paid to send the one faculty member with some background in the field to the ATTW conference to gather ideas. That faculty member came back armed with the realization that professionalizing our courses meant more than teaching memo-writing; if we were serious, we had to come up with a way to incorporate some of the theoretical and pedagogical background that constituted a meaningful professional writing program.

There was one big problem: we had neither the fiscal resources nor the student population to consider a separate program. Nor did we necessarily want to. We knew that other departments were undergoing similar struggles to combine their traditional offerings in the major with more skills-based courses—the Art Department, for example, had added Computer Design I and II. And the Computer Science Department was working on an emphasis in software design, and wishing that someone on campus taught technical writing. The chair of the department mentioned this to the Dean, who relayed the good news that the English Department was adding just such a course. She discussed with Prof. Hathaway her desire to require that software engineering students take it even though it was an "English Major" course, and the two began wondering what else they could combine. In a strange confluence of need, therefore, members of faculty from several departments looked around and saw that a shift was occurring and that the best way to capitalize on it was to combine forces.

From here, Prof. Hathaway asked the Dean if he could convene an interdisciplinary committee to come up with a program design, knowing the strain on resources that creating a new minor can put on departments: hiring new faculty, adding new courses, updating facilities. Creating an interdisciplinary minor seemed to be the best option: it would give students some substantive breadth and depth to preparation for the professional workplace, *and* it would spread the responsibility for this program among a number of participating departments, so as not to overly burden or alter the curricula of any one department. We believed we could answer the call for professionalization by adding several new courses in various disciplines and pooling existing departmental resources.

Thus, in the Fall of 2000, we convened a Technical and Professional Skills Committee, with representatives from the Computer Science, Communications/ Journalism, and English departments. As the committee considered the objectives and prospective course inclusions for such a program, we invited in the Art and Speech departments as well. In our deliberations over that year, we reviewed the various technical and professional communication programs in the region and studied student need to better determine the service and draw of such a program.

We followed up by designing an eighteen-hour program that had a six-hour core of two 100/200 level courses: Technical/Professional Writing I from the English Department and either Overview of Computer Science or Business Computer Systems, which students in the College of Business could take in lieu of Computer Science. The other available courses provide the minor with the variety the Committee was hoping for, with offerings from the following departments:

- Art Department: Computer Design I & II
- Communications/Journalism Department: Advertising Copy Writing, Feature Writing, Writing for Broadcast Media, and News Writing
- History and Philosophy Department: Ethical Issues and Computer Technology
- English Department: Technical/Professional Communication II
- Computer Science: Web Programming
- Speech/Theater Department: special topics course in either Communication in Training or Organizational Communication

To insure that students in the program would benefit from the variety of course offerings and not end up taking a facsimile of a participating department's existing minor program, we stipulated that, excepting internships, no more than two courses from any one department could count for minor credit. This allows students who might be interested in graphic design to focus more computer courses, for example, while allowing those interested in Communication to create a "different" minor with courses more suitable for them, selecting two courses from the Communications/Journalism Department and two from somewhere else. Thus, the program we envisioned and eventually put into place is more comprehensive and diverse than almost any program in the area. And though creating a multidisciplinary program was logistically our best option, it has provided us with some unforeseen pedagogical benefits as well.

The Multidisciplinary Edge

Financial considerations were one of those pragmatic realities that we faced when we set about creating the new minor. When developing the program, we were very mindful of finite resources on campus and other concerns that the Chancellor's Office might raise down the road, and so we worked to allay fears. For example, recent university-wide technological initiatives meant that regular technology and software updates kept departmental computer labs and equipment up to par; thus we did not have to ask for any start-up money. The Communications/Journalism and Art departments not only upgraded but also expanded their computer labs, which helped make space for students in the minor. To insure that there would be available space in the participating departments without the burden of new sections, we proposed an initial program cap of thirty students; in fact, we did a seating-availability breakdown, course by course.

But spreading the burden of a new minor to various departments has sound pedagogical as well as resource benefits. Allowing only two courses from any one department to count for minor credit, for example, certainly eases potential department overload. But the benefits our graduates will get from a multidisciplinary approach, by exposing them to fields other than their own, is even more important. Few English majors will take a computer science course, the other core course in the minor, if they don't have to. Some classes in the Art and Communication/Journalism departments are closed to students outside of the major or minor, so without declaring this minor, students would not be able to take certain courses. Even if our university was large enough to have a technical writing minor track in the English department, these types of courses would not be available to them.

And exposure to a variety of courses in other fields is vital for technical communicators, since it is so often their job to serve as a liaison in the workplace. Whiteside's survey of recent technical communication graduates and their managers in the field is applicable. While there were some differences between what students perceive they need and what managers wish they had (like learning computer software and languages), there was strong agreement that students need more preparation in the following four areas: business operations, project management, problem-solving skills, and scientific and technical knowledge (313). It is our hope that students get some preparation in these areas during the time they are in the minor, working as they will in different departments. Writing students with a knowledge of business systems, which they can receive in our minor, will be able to understand more completely the role of technical communicators in the workplace, a need that many managers in Whiteside's survey noted. In addition, we saw the opportunity to expand our students' writing, judgment, and speaking skills by including courses such as advertising copy writing, ethical issues and computer technology, and topics in organizational communication. This range of skills is not often emphasized in technical programs, yet it reflects a broader preparation that is very applicable in the professional workplace.

Another example of the unforeseen benefits of the multidisciplinary minor concerns the directorship of the minor. Because the program is not housed in any one department, the potential burden of directorship will not be limited to one department, where course-release for the director might cause hardship. But this has an additional benefit of reminding the students (and the university) of the true multidisciplinary nature of the minor. Though the English Department was one of the departments that spearheaded the proposal, we did not have the intention or the desire for it to be particularly our minor. There is question as to whether even a "regular" technical writing or communication program should be housed in the English Departments of universities, as MacNealy and Heaton describe in "Can this Marriage be Saved?" Their survey questions where the right home for such programs is, based on the difficulties that some technical communication faculty members have in English departments, where they get neither respect nor support from their colleagues. One of the solutions these authors propose is to make the program interdisciplinary, a choice "which would seem to suit a large group of respondents in our survey" (58). Thus, while a revolving directorship serves the practical function of lessening possible strain on departments, it also insures the long-term interdisciplinary character and ownership of the program.

Our proposal for the interdisciplinary Technical/Professional Communications Minor passed university review in Fall 2001. However, it sat unresponded to for a year and a half in our State System's Chancellor's office. When it was finally readdressed, we took the opportunity to update the information in it and expand on our vision of its apparatus. Finally, in the Spring of 2003, we were given approval to begin implementing and advertising the minor.

The next section, focusing specifically on revisions to our original proposal, describes the overall rationale for creating a totally new professional communications minor. It discusses the strategies we used to support our claim for the need for this type of program, through analyzing both other colleges' programs and student need, and our plan for the program's assessment.

PROGRAM RATIONALE

In our original proposal from the Fall of 1999, we completed a survey of area colleges (within a one hundred mile radius) to see what kinds of courses or programs in technical communication they had. One of our concerns was that after two years, our rationale for the program's importance and the area's need for it was a little outdated. We set out to prove our minor was now even more in demand.

Showing Need for the Program

In updating our proposal, we expanded our hunt for technical communication programs to all schools in Pennsylvania and in the Baltimore area. We discovered that several area colleges were now offering either tracks within a major or an actual degree in Technical, Professional, or Business Writing, a fact that we found both welcome and alarming. On the one hand, the increase in technical communication programs implied a need for graduates of such programs in the workforce and lent support to our assertions; on the other hand, it meant that we needed to get our own program going, so as to not get left behind.

But we were able to draw two other conclusions from our survey that were more satisfying. First, while a number of State System schools have similar tracks or minors, none are located in South Central Pennsylvania. Thus, within our own system of universities, this minor would fill an area need. And second, while a few schools had an interdisciplinary focus to their tracks, our range of multi-departmental offerings made our program unique. This could potentially draw students to our school who might have otherwise attended a different one in the system.

Another way we showed the importance of our program was to relate the growth of the technical communications field to our Chancellor's Office in a meaningful way. We knew that jobs in the field were on the rise; we weren't sure they knew. So we passed on the fact that the Bureau of Labor Statistics anticipates that opportunities for technical writers are excellent: "Demand for technical writers and writers with expertise in specialty areas, such as law, medicine, or economics, is expected to increase because of the continuing expansion of scientific and technical information and the need to communicate it to others" (Bureau of Labor Statistics Job Outlook).

To give this data a local slant, we pointed out the increased enrollment in area chapters of the Society for Technical Communicators (STC), a non-profit group dedicated to highlighting the work of technical communicators and the largest of its kind. Even more exciting was that a whole new chapter of the STC centered in the Harrisburg area, the Susquehanna Valley chapter, had been created in 2000. Like in most chapters, members include both professionals in academia and professionals out in the field. Its thriving status speaks to the large presence of technical communicators in our area and their desire to participate meaningfully in their careers and in their community. We plan to have at least one member of our Advisory Committee members (see section IV) join this chapter, helping foster connections between the university and area companies.

We also highlighted the excellent salary opportunities in the technical communications field, based on the survey of national salaries that STC performs every year. By comparing the salaries for 2000 to those in 2003, we were able to show that salaries were not only highly competitive but that they continued to rise. We also noted STC's assertion that there existed a very small difference in pay between men and women—women can earn 97% of their male counterparts. Judith Glick-Smith, the 2002-2003 past president of STC, writes that "this smaller 'gender gap' points to financial opportunities for women in the growing field of technical communication" ("Salary Survey"). This fact is significant because one of the things the Chancellor's office wanted to see was how the minor would attract women, minority students, and non-traditional students. We were proud that we could provide such a telling statistic.

Lastly, we turned that knowledge into a more personal recommendation, interviewing Shippensburg's Director of the Career Development Program, Dan Hylton. He reiterated for the Chancellor's Office what our research showed:

> For the past several years, the National Association of Colleges and Employers (NACE) has published research indicating that communication skills top the "Perfect Candidate" list of desired qualities in candidates being interviewed for entry level professional positions. In particular, technical writing and computer literacy are highly sought in new hires. [...] From our campus and job fair feedback to national trends and research, we would strongly support the implementation of a technical/ professional communications minor at Shippensburg University.

Having researched the need for a program like the one we envisioned in the area and in our own university, we felt confident that the new minor would succeed.

Attracting Students to the Minor

Another item the Vice-Chancellor wanted to see was how we would attract students to the program, especially minority students, nontraditional students, and undeclared students. We contacted our Dean of Admissions and our Dean of Undeclared Students to see what they thought. Both were highly supportive, recognizing that the combination of computer and business skills with writing and communication skills created a highly desirable program of study. Joseph Cretella, Director of Admissions, agreed that prospective students might be attracted to the minor, something we had hoped would be the case: that its uniqueness might actually draw students to Shippensburg instead of to another college in the State System. He writes:

> I do believe the combination of technical writing and computer skills could be the key to elevating the interest in the program. We get a ton of interest in computer design programs. Instead of computer graphics with a huge

amount of math and programming, your proposal could be an alternative, which would fine tune their writing skills supported by the computer.

Dr. Marian Schultz, Dean of SU Academic Programs and Services, who works with both undeclared majors and minority students, was very enthusiastic. In discussing the minor's appeal to students who enter the university undeclared, she said:

> Philosophically, we encourage undeclared students to see their college degree more broadly and to select academic programs and experiences that will help them develop lifelong skills that are transferable to many occupations. Primary among these are good oral and written communication skills, as well as technical proficiency with various computer programs and applications. [. . .] The proposed Technical/Professional Communications Minor, which helps students develop these skills [. . .] will provide them with a competitive edge in the work place.

She also thought the program would be attractive to students of color, many of whom are enrolled in business programs, as undeclared, and in communications majors. She writes, "The Technical/Professional Communications Minor will provide our students of color with the opportunity to enhance their educational experience and to increase their employability by helping them to develop the communication skills employers are looking for." As the minor grows, the director will make a strong effort to meet with leaders of programs like Ethnic Studies, Multicultural Student Affairs, and Women's Studies to discuss ways to best attract a diverse student population to this minor.

Lastly, in keeping with other minors on campus, the program has no grade-point average requirements, so that any student who is attracted to the minor can take it. While it might be tempting to use the popularity of this minor to set minimum entrance requirements and attract stronger students, to do so could box out some students who might particularly benefit from the professional knowledge and skills afforded by this program, in particular these non-traditional or under-prepared populations.

When the minor was approved, we sent our promotional flier to all academic advisors prior to scheduling, and Prof. Hathaway visited with department chairs at our College Council meetings. We also put an article in our student newspaper and in the undeclared majors' newsletter to let students know that the new minor had arrived.

Assessment Tools

A final concern of the Provost and Chancellor was our assessment strategy. Our original proposal did not outline an assessment plan in any detail; when we went about revising it, therefore, we took more time thinking about what goals we wanted to accomplish with the minor and how we might design an assessment strategy to meet those goals.

In our case, we had an additional difficulty in trying to develop an assessment plan for a program that hadn't even been implemented yet. But we knew that a well-thought out assessment plan was invaluable for meeting our program's goals. As Jo Allen suggests, assessment can be "powerfully effective for planning, designing, and promoting distinctive programs and then recruiting desirable students and faculty"(93). Further, we agreed with our Provost that a solid assessment plan was especially vital for an interdisciplinary minor, which lacked the curricular structures found in single-major programs.

But first, we needed to clarify what we wanted our assessment to do. Jo Allen echoes the types of questions we were asking ourselves: what do we want to accomplish with the assessment, and how will the information we receive from the assessment be used (98)? The committee had a clear sense that it wanted to focus on student-outcomes assessment, looking at what students learned from being in the minor. We thus developed goals and objectives of the assessment plan:

- To measure student learning and skills as appropriate for professional and post-graduate educational opportunities
- To review student attitudes and feedback about strengths and weaknesses of the program experience
- To solicit program alumni feedback
- To review appropriate design and effectiveness of constituent courses
- Based on the above, to make periodic program changes as deemed appropriate

Similarly, we needed to better clarify how our program assessment would benefit from the evaluation criteria we had decided upon. We had briefly listed some evaluation tools in our initial proposal—course evaluations, student portfolios, exit interviews, and alumni surveys—but we hadn't really thought about what we were looking for with all this information. Thus, when we went back to flesh out our proposal, we began with some goal statements: what did we want the students to know when they completed the minor? What skills did we deem most important?

We outlined four basic skills we wanted out students to graduate with:

- Shows facility with appropriate computer applications
- Has knowledge of and can apply the conventions of professional writing
- Presents material in an organized, clear fashion
- Demonstrates critical thinking appropriate for professional tasks

These are clearly broad, broader than what is listed in most programs' skills assessment. But with such a range of courses in the minor, and the various combinations of courses that each student could take, we felt it necessary to think as broadly as possible. Nonetheless, these items can be used by each department or course instructor as meaningful indicies of student growth and preparation.

From there we felt we had a better sense of how the tools we would use to measure these skills would help us monitor the program and evaluate its effectiveness, based on a student-outcomes assessment. These tools consist of a portfolio system, samples of which would be reviewed every other year, and both an exit questionnaire, taken by students in non-core courses, and an Alumni questionnaire, surveying graduates two and five years after their graduations.

Portfolios form the basis of our assessment, as they serve so many vital functions: they provide us with a body of work from both core and elective courses; they provide a way to sample the actual work students are doing, and they give students some control over the work by which they choose to be evaluated (though when we use the portfolios for assessment purposes the students' names will not be provided). Since having experience with appropriate computer applications is one of our skill outcomes, in addition to the traditional writing skills, we want to collect both electronic and written projects for the portfolio. Each portfolio will contain work from both of the core courses, Tech Writing I and either CMP 102 or BUS 141, and then two additional projects from other courses, one of which must be 300-400 level or an internship.

Figure 1 shows an edited copy of the assessment rating form (on the actual form, all courses are listed), which clearly lists our goals for the students and the rating scale. Faculty in pertinent courses will make anonymous numbered, dated copies of designated minor students' projects and turn them in each semester to the program director.

We also plan on administering two sets of questionnaires, separate from the university's standard course evaluations. Both the Exit and Alumni questionnaires will be written and will use the same numerical scale as the portfolios. The categories we plan to cover include: the program overall, the students' individual courses, and the students' own assessment of their professional knowledge and skills. For the Alumni questionnaire, we will add a place where the students can update us on any professional or post-graduate education. These questionnaires are currently being developed, as we graduate our first batch of students who have completed the minor this December.

Student Year _____   Student No._____

	Shows facility with appropriate computer applications	Has knowledge of and can apply the conventions of tech writing	Presents materials in an organized, clear fashion	Demonstrates critical thinking appropriate for professional tasks
ENG 238				
CPS 103				
ART 217				
(list continues for all courses)				
Internship				

Scale: 1, excellent; 2, good; 3, adequate; 4, unsatisfactory

FIGURE I. TECHNICAL/PROFESSIONAL COMMUNICATION MINOR PORTFOLIO ASSESSMENT RATING FORM

After fine-tuning our assessment plan, we feel that we have a good blueprint to help us begin to evaluate our program and how the minor will contribute to our students' career possibilities. The next section discusses the program's administration and its immediate success with the students, verifying the confidence we had about the program and its attractiveness to students from across the college.

PROGRAM ORGANIZATION AND ADMINISTRATION

Once approval to begin was in hand, we set about preparing to inaugurate the minor program. Several concerns faced us that had not been part of our original plan. First was the changed state of availability in participating

(more)

classes. The university had admitted record-size entering classes for two years running since 2001; overall enrollment, including increased transfers, was up several hundred. Some of the participating courses, such as Technical/Professional Writing I and Computer Design I, had experienced unforeseen demand and were turning some students away every semester. This meant that seniors and juniors were filling most if not all of the available seats and that some who would want to declare the minor might not have enough credit hours remaining to take it. In addition, we knew, anecdotally, that the word of pending T/PC Minor approval had already generated considerable student and faculty interest.

We decided to pursue two solutions: first, to make passing the core courses a prerequisite to declaring the minor, which would require students seriously interested in the minor to establish their eligibility; and second, to try making a number of summer sections available online, particularly those for the core courses. The College of Business and the English and Communications/Journalism departments took up the online invitation; in fact, the request from the minor program gave them the final nudge in a direction already being seriously considered. Thus, seven of the thirteen courses in the minor were offered this past summer, and five of those, including the two core courses, were offered online, making our capacity for enrollment in the minor potentially unlimited, without stretching existing resources.

Once we lined up our courses and had the online offerings arranged, we began advertising. In early October of 2003 we sent out our promotional flier to all faculty and academic departments. We also sent a flier and a note to all enrolled students who had already taken one or more of the core courses and who had sufficient credit hours remaining to undertake and finish the minor in time to graduate. Our efforts paid off, as we had expected. Advising for spring semester began mid-October of 2003, and as of mid-November, we already had thirteen students registered for the minor with additional students signing up daily for advising conferences. This substantial early response clearly indicates that our new minor will be a highly attractive offering to students from across the university.

Looking Ahead

To fully complete the goals we set out for ourselves in our proposal, we still need to finalize several tasks within the next year. One of the first items on our list is to create and convene our advisory board, composed of area professionals, faculty, and students. We plan to use this body to help us to evaluate and assess the program, but particularly, to help develop off-campus internship opportunities and aid in career planning. To that end, we have scheduled meetings with the College of Business to discuss promoting the minor and assembling business professionals in the surrounding Carlisle-Harrisburg areas.

Our second major task is to begin initiating our assessment plan cycle. When we revised our proposal and included a detailed assessment strategy, we included a timetable for when those instruments would be used. All of the assessments will be performed by the Program Director in conjunction with the advisory board.

Below are the assessment instruments and our planned timetable for their implementation.

1. Portfolios

    Written and electronic student portfolios of key course projects:
    - one each from each of the two core courses;
    - two additional from other courses, one of which is 300-400 level or an internship.

    Portfolio folders, record sheets, and student consent forms:
    - to be set up, signed, and explained by program director when students sign up for minor;
    - to be shared with participating faculty.

    Completed portfolios to be reviewed:
    - starting third year of program;
    - every other fall, from a significant sample of program students completing requirements the previous two years

2. Exit Questionnaire

    Anonymous exit questionnaires:
    - one for each student completing requirements, in his or her last program semester;
    - to be distributed by professor in all non-core courses near the conclusion of each semester.

    Exit questionnaires to be reviewed:
    - starting third year of program;
    - every other fall.

3. Alumni Questionnaire/Survey

    Periodic alumni questionnaires:
    - one for each program alumna/alumnus of a given year;
    - to be mailed on the second and fifth anniversary of SU graduation;
    - Second anniversary mailings in spring of even years;
    - Fifth year anniversary mailings in spring of odd years, by program director.

    Alumni questionnaires to be reviewed:

- starting fifth year of program;
- every other fall.

4. <u>Syllabi Review</u>

All participating courses' syllabi/descriptions for the most recent year to be reviewed:
- starting third year of the program;
- every other spring.

We also needed to establish our program's publications, a newsletter and a website in support of the program. The website was completed this past fall, as was an informational brochure to be placed in the admissions office. We hope that either a graduate assistant or an intern can be regularly appointed to help with a newsletter and other program maintenance. In addition, we want to begin implementing several other services in support of the minor. First, we plan on instituting annual job fairs and resume/professional portfolio workshops. We also want to continue to improve our marketing of the minor to incoming and prospective students by promoting the minor during freshman orientation and by holding open houses.

Lastly, we will begin reviewing other courses for possible inclusion in the minor. We initially developed the minor's course offerings based on the strengths that our faculty had and by choosing classes that were already available. We foresee that a number of departments will wish to contribute courses to the minor or to increase the number of courses they currently supply. Also, if demand for the program continues to rise, it is possible, based on exit and alumni questionnaires, that new courses can be created to fill certain gaps or student desires. This will continue to provide our students with the interdisciplinary focus we want the program to supply.

CONCLUSION

To meet increasing student interest in technical communication opportunities, faculty from a number of disciplines met and developed a unique multidisciplinary program, bringing together a variety of courses from across the college curriculum and organizing them into a flexible and substantive minor. The Technical/Professional Communications Minor will allow participating departments to better integrate shared expertise and to better utilize existing resources, with a minimum of additional resource needs. Most importantly, the skills and knowledge available in this program will offer a synthesis of academic and professional preparation that truly reflects the liberal arts core of Shippensburg University. We are particularly proud of our ability to integrate, rather than segregate, professional skills and traditional higher-education skills, and meet in the best way possible the whole of the university's mission.

WORKS CITED

Allen, Jo. "The Impact of Student Learning Outcomes Assessment on Technical and Professional Communication and Programs." *Technical Communication Quarterly* 13.1 (Winter 2004): 93-108.

Bureau of Labor Statistics Job Outlook. 2003. Bureau of Labor Statistics. United States Department of Labor. 22 July 2004. http://www.bls.gov/oco/ocos089.htm

Cretella, Joseph. Email Interview. 18 April, 2003.

Dan Hylton. Email Interview. 18 April, 2003.

Johnson-Eilola, Johndan. "Relocating the Value of Work: Technical Communication in a Post-Industrial Age." *Technical Communication Quarterly* 5.3 (Summer 1996): n.p. Academic Search Premiere. EBSCOHost. Shippensburg University Library. 5 July 2004.

MacNealy, Mary Sue, and Leon B. Heaton. "Can This Marriage Be Saved: Is an English Department a Good Home for Technical Communication?" *Journal of Technical Writing and Communication* 29.1 (1999): 41-65.

Rude, Carolyn, and Kelli Cargile Cook. "The Academic Job Market in Technical Communication, 2003-2004." *Technical Communication Quarterly* 13.1 (Winter 2004): 49-71.

"Salaries Rise for Technical Writers & Editors." October 2003. Society for Technical Communication. April 2003. http://www.stc.org/pressrelease_srise.asp

"Salary Survey 2003." Society for Technical Communication. April 2003. http://www.stc.org

Schultz Marian. Email Interview. April 16, 2003.

Whiteside, Aimee. "The Skills That Technical Communicators Need: An Investigation of Technical Communication Graduates, Managers, and Curricula." *Journal of Technical Writing and Communication* 33.4 (2003): 303-318.

10 Reinventing Audience through Distance

Jude Edminster
Andrew Mara

The following describes our attempt to extend and resituate the graduate programs in Scientific and Technical Communication at Bowling Green State University. Our revisioning of our program includes developing an online graduate certificate program in international technical communication.

OVERCOMING DISCIPLINARITY

One of the challenges in offering a certificate program in Northwest Ohio comes from the lack of high-technology jobs in the area. Although online programs ostensibly hold the promise of erasing distance limitations altogether, there still seems to be a general rule that people will look for programs in their area. This is not surprising, considering that people who work in technology centers like Austin, Palo Alto, and the Research Triangle have likely heard little to nothing about Bowling Green State University. So, with few technology workers to train, our S & TC program wanted to offer something more recognizable to the business writers that physically and professionally inhabit the area from Columbus to Detroit, an area that roughly corresponds with the I-75 corridor. Thus, our challenge included thinking of courses that would apply more generally to professional writers of all stripes. Because our goal was a graduate certificate program, and not a full-blown online graduate degree (we already had a resident graduate program in S & TC), it was easier to create a more generalized program.

Although the strategy of adding a certificate program may seem either an overt attempt at pure growth, or even a step back from our resident undergraduate and graduate degree programs, it really has had more of a leavening effect. By demonstrating that a program can occur exclusively online, and by demonstrating its value to the enterprise community, our program can make a case that future expansion and curricular realignment can and should take these two approaches. Fortunately for us, this also helps the S & TC program integrate two new professors who both have expertise and experience in both business and online issues and practices.

Having experience and expertise are not enough in the competitive forum of acquiring academic resources, though; Bowling Green State University has programs in both Communications and Visual Communications Technology which have also been eyeing the creation of writing programs that serve the business community. Fortunately for us, the English Department has been at the front of efforts to offer online courses. That fact, coupled with our program's reputation for more cutting-edge approaches within the department, gave us some leeway in creating an online certificate.

NEW FACULTY/NEW DIRECTIONS

The use of an online certificate to develop our Scientific and Technical Communication program is one that combines institutional and personal history. As new members of our English Department, both of us were immediately thrust into the formidable job of heading up our program (Dr. Edminster in her third year and Dr. Mara in his second). Like many other institutions shifting their curricular focus to digital approaches (University of Central Florida's "Text and Technologies" program, North Carolina State's PhD in "Communication, Rhetoric, and Digital Media," and Michigan State University's "Rhetoric and Writing" program just to name but three), Bowling Green State University's English department has been enjoined to find ways to offer more courses through online distance education and hybrid computer classes with online components. In point of fact, our Dean for Continuing Education urged our department to take "bold measures" (a prompting that has led our department to offer more online courses than any other academic department).

BGSU's English Department hired both authors in the midst of the increased academic pressure to offer more online and distance courses. Importantly, both new S & TC professors have an expertise in electronic communication (Edminster in electronic theses and dissertations and Mara in hypertext rhetoric and histories). While these hirings may be nothing more than coincidence, they offered a kairotic opening to propose the certificate and to shift our programmatic focus from more traditional technical communications concerns (rhetorical approaches, documentation) to some newer issues (technical marketing, entrepreneurship, and globalization). Issues such as audience, departmental territory, resource allocation, and finding personnel to teach courses were all complicating factors that needed addressing before the certificate was proposed. Proposing the certificate introduced complications as well as providing a certain amount of institutional cover to solve these problems because of the promise of online course delivery. As more experienced faculty well know, proposing untested solutions to enduring problems like scarce resources can be a dangerous business. A new faculty member might propose a bold solution; two new faculty can get downright giddy when looking for possible solutions.

More than just an empire-building opportunity, this certificate program has proved advantageous in helping the English Department bridge the gap between different programs. In the process of formulating the new certificate program, we serendipitously connected with another member of the faculty who also specializes in digital issues, Dr. Kristine Blair. Through our discussions, we quickly ascertained that there was much common ground and many similar aspirations percolating in the English Department. Our literature program is reorganizing around a textual studies paradigm and our Rhetoric and Writing program has increasingly integrated online components like Electronic Dissertations and eportfolios. All of these trends indicated a migration towards a more digitally-oriented curriculum—a migration that a Scientific and Technical Communication graduate certificate could propel. Working in concert with the Rhetoric and Writing Program, and by including a proposal to help all English faculty develop online courses, Drs. Edminster, Mara, and Blair all applied for an Ohio Learning Network grant for a "Digital Studio." Using a learning community model, we carved a conceptual space to evaluate, and even migrate some of our curriculum towards an online environment. The studio repurposed then-unused offices and computer labs to host group workshops, one-to-one training forums, and curriculum revision document drafting in order to help stimulate discussions and eventually to help drive curriculum change. The Ohio Learning Network "Digital Studio" grant allowed the three co-Primary Investigators to hire a small staff and fence off time that would otherwise be taken up with teaching, and granted a certain credibility in justifying our declaration of off-hour facilities as studio space. In short, the grant allowed us the *ethos* to structure an otherwise unauthorized environment. By re-mapping idle classrooms and offices as studio spaces and calling meetings to fill those spaces with digital training or future planning, two non-tenured faculty members were able to triangulate several programmatic needs in the department and eventually create a new programmatic iteration.

GOING DIGITAL/RECREATING CONTEXT

While English departments have traditionally deployed a range of common objects, referents and places to define what we do (the classroom, the novel,

Shakespeare, Nineteenth-Century American Literature), many of these referents take on a different embodiment when they go digital. While English Departments have been entirely comfortable disputing which referent should have precedence, there has been a tangible avoidance of discarding these common referents in all but a few universities. Technical and Scientific Communication programs are no different. Stuart Selber's and Johndan Johnson-Eiola's recent compilation, *Central Works in Technical Communication* signals a traditional gesture of binding departments, programs, and academic disciplines together with artifacts like editions and anthologies. While this is neither surprising nor alarming, it does signal a sort of disciplinary inertia that we have to face when proposing something more negotiated and ephemeral, like online transactions, as a centrally-organizing principle of a program. In the case of our program, the organizational principle—creating a mutually-negotiated space—involves emphasizing the ground upon which these negotiations can occur.

Much recent discussion has ensued on the role of different types of electronic contexts and contextual aids. Websites, MUDs, MOOs, blogs, WIKIs, and other hybrid electronic fora have all provided fodder for journal, conference, and even blog discussion. Instead of positing which form best facilitates student interaction and learning (along with the multiple political connotations), our program has decided to emphasize the importance of providing the ground and emphasizing the constructedness of both place and agreement. The negotiations will include not only approaches to coursework and pedagogy, as were mentioned earlier, they will also include notions of techne, technology, and even evaluation. Program evaluation provides the tool through which traditional notions of success will be challenged.

Using BGSU's Provost-championed University Academic Plan (*Inquiry, Engagement,* and *Achievement*) as both exigence and starting point allowed us an additional measure of freedom in establishing the program. In the Academic Plan, Bowling Green State University emphasizes the new ways to integrate online technology as part of the university learning experience—in fact, the push towards a vaguely defined "New Media and Emerging Technologies":

> By providing focus to the current activity and a platform for investigation into emerging technologies, we can achieve national prominence in creative teaching, innovative research, and industrial collaboration. (Plan 19)

The push towards new media (officially one of five "themes") granted potential power to two relatively powerless faculty members (at least in the hierarchy of tenure-line faculty members and administrators). Realigning our program with a broader, university-wide planning document allowed two young professors to create a future vision that might help staff, senior faculty, and even administrators make future requests for funds (or at the very least, request smaller cuts in funding).

Seizing upon the general impetus of building a technologically savvy university, we were empowered by the dean of Continuing Education to craft our online certificate. As a way to build upon an already large commitment to online curricula, our certificate presents a wider strategy to migrate online en masse. Also voiced in the Academic Plan is the university's renewed commitment "to develop students who seek intercultural and international engagement and who possess a capacity to relate to diverse others at home and abroad." The international focus of the certificate will support this institutional commitment as well.

Further support for the graduate certificate's international focus is contextually situated in the university's exchange program agreement with Xi'an Foreign Languages University (XFLU) in China. Under this agreement, Bowling Green State University has the flexibility of sending either two or three faculty members or graduate students (holding an M.A.) to Xi'an Foreign Languages University in exchange for three XFLU graduate students who come to Bowling Green. Our Scientific and Technical Communication MA Program regularly accepts and enrolls exchangees from XFLU under this agreement, and S & TC faculty have also participated in the exchange, teaching courses in technical writing at Xi'an and working with Chinese graduates of our MA program to try and establish a technical writing program at XFLU. As Ping Duan and Weiping Gu point out in their article, "The Development of Technical Communication in China's Universities" (434) technical writing, as a subject of study, is virtually non-existent in China. At least two of our Chinese MA students have produced master's theses documenting the current need for technical writing programs in China and exploring the feasibility of establishing such programs at XFLU and other, more technically focused universities. As the need for such programs to support China's rapidly expanding technological and global economic development continues to grow, an online graduate certificate program such as the one we have developed may appear extremely attractive to a variety of Chinese businesses and industries interested in developing employee communication skills as a long-term, quality management strategy.

When Edminster arrived as a new hire in the S & TC Program in the fall of 2002, tentative plans for the online graduate certificate program were already germinating in the mind of the current (and veteran) S & TC Program director. The desire to create the certificate arose, in part, out of the program director's desire to move S & TC from the confines of the English Department. A year or two earlier, the director had tried to convince the Dean of Arts and

Sciences and the Provost that S & TC belonged in the College of Technology rather than in the College of Arts and Sciences as part of the English Department. The administers were not persuaded, so the director calculated a different, less dramatic move in the direction of independence—a collaboration with the Continuing Education Program to develop and market an online graduate certificate program. This collaboration coincided with the university's new initiative—developing and offering more online courses campus-wide.

However, plans for the certificate were embryonic, requiring a good deal of brainstorming and late afternoon coffee as Dr. Edminster and the former program director began to target and assess the needs of a market, to conceptualize courses that would meet those needs, and to negotiate with an associate dean of Continuing Education, who is also a member of the English Department. Many of the potential students for this program are in industry seeking ways to increase job security or to improve their abilities to communicate in an expanded, global market, but lack the flexibility of being able to come to campus for primarily daytime classes. Others reside in other countries, either in residential programs in those countries (including those who teach and/or study at Xi'an) or also working in business and industry without the flexibility of leaving jobs and countries to enroll in residential programs. Still others are seeking upgrades or updating (certification) of skills and knowledge without need for a degree. The program is designed so that students who are enrolled in a degree program can substitute some courses for residential requirements for the degree, while others can pursue the entire certificate program.

Gradually, student performance objectives for the program began to take shape. After completing the program, students were supposed to possess the ability to:

- Create common technical vocabularies within "transaction" cultures as they are socially constructed through intercultural interaction
- Analyze cultural bias
- Employ forms of project management that facilitate intercultural collaborative writing (dialogic rather than hierarchical)
- Apply "learning organization" management concepts in order to learn from diversity
- Develop collaborative relationships that generate mutual knowledge within "transaction" cultures
- Assess and reflect on their collaborative processes
- Design effective usability tests for documents with international or culture-specific audiences
- Develop evaluation criteria for processes and products

- Analyze processes by which specific cultural values interact

Courses eventually included: (1) Technical Communication for a Global Marketplace, (2) Technical Editing for a Global Marketplace (3) Research in International Technical Communication, (4) Ethics in International Technical Communication, (5) Visual Displays of Information for International Audiences.

In our field, international technical communication has traditionally focused on developing both the awareness and the skills necessary to understand how cultural difference affects communication in various technical contexts and to plan for and design documents that meet the needs of an audience that is both culturally diverse and culturally specific. This certificate program relies to some extent on this traditional approach by teaching such skills as (1) how to analyze cultural bias, (2) how to analyze international and nationally specific audiences, (3) how to design effective usability tests for documents with international or culture-specific audiences, and (4) how to translate the culture as well as the language of technical documents. In addition, the program places significant emphasis upon the growing awareness within technical communication research that the application of static notions about particular cultures on the part of technical communicators can degenerate into the reinscription of cultural stereotypes that obstruct communication rather than facilitate it. Thus, our certificate program also emphasizes the need for technical communicators to understand that: (1) every communication situation is context-specific, (2) although context includes culture, cultures do not communicate with each other, individuals do, and (3) the culture that defines individual international communication situations is a "hybrid" or "transaction" culture, which is constructed by the participants as they interact and negotiate their cultural differences. The certificate is designed to prepare students to function in the global workplace by instructing them in how to apply both knowledge about culture and knowledge about negotiating cultural difference in individual communication contexts. The online certificate will situate communicators from different cultures in a mutually constructed third space via the use of online discussion boards and virtual classroom courseware tools. This hybrid space cannot be mapped out in advance of the communication situation, but instead must be negotiated with the paralogic hermeneutic approach described below.

PARALOGIC HERMENEUTICS

As we sketched out the specifics of the certificate building process (some of which occurred before we arrived at BGSU) we found it useful to employ

Thomas Kent's theory of paralogic hermeneutics as a lens through which to view negotiations between the Department of English and the Department of Continuing Education, which oversees all online course development and marketing at our institution. In his work *Paralogic Rhetoric*, Kent describes a communicative approach that fits well in what Bill Readings might have affectionately called the postcultural university. Kent recognizes the need to serve a wide range of students who do not necessarily buy into a liberal arts or strictly cultural education; he forwards the proposal to refocus education on contextual practice:

> Paralogy is the feature of language-in-use that accounts for successful communicative interaction. More specifically, paralogy refers to the uncodifiable moves we make when we communicate with others, and ontologically, the term describes the unpredictable, elusive, and tenuous decisions or strategies we employ when we actually put language to use. (3)

The uncodifiable moves make it impossible to definitively map a curricular path for students (especially considering that we have an active international student population and are seeking to increase our focus on international/post-national communication issues). Instead, our certificate program seeks to acknowledge and facilitate the future interactions/decisions/strategies that our students will inevitably take throughout their time in our program.

In order to help our students articulate the contingent and increasingly-diverse communications scenarios that they will face in trans-national, international, and even intra-national situations, we have shifted our focus from the more concrete matters of "content" to include the more ephemeral matters of setting (online) and even unpredictability as an (un)grounding feature. We are not making radical shifts into relativity, however; instead, our certificate program seeks to distribute responsibility across the range of participants. Kent calls this process of distributed responsibility triangulation:

> In order to surmise if our marks and noises create any effect in the world, we require at least an-other language user and objects in the world that we know we share. In order to communicate, we need to triangulate (90).

Kent's use of the term "triangulation" refers to the interaction between two communicators and the world they share. Other language users and worldly objects take on a different dimension in an online curriculum, especially insofar as we cannot count on common facilities and classroom locations to create boundaries and occurrences as our objects of commonality.

At the same time, this paralogic challenge helps our program enact workplace strategies that students will employ in their future workplace settings. Technical writers increasingly have to face writing situations where they will likely never meet or physically interact with their users. Internal documentation increasingly gets fed into websites that will be read and implemented in different countries. External documentation is likewise sent to multiple places and audiences who never physically intersect (often translated by localizers to match an even greater array of paralogic, triangulated interactions).

TRIANGULATING IN A THIRD CULTURAL SPACE

This triangulation provides us with a guiding tool for constructing a new certificate. Classes will include few "plug-and-chug" formulas that give the impression that audiences are unchanging, workplaces stable, and technologies separate from interaction. The pedagogy will integrate the back-and-forth motion between students in group interactions. If triangulation provides a good guiding tool for course construction, it provides an even more powerful heuristic to dealing with the complications we face as new faculty, new program directors, and new technologists.

In order to help students triangulate, we are attempting to simultaneously empower them to participate in meaning-making and to recognize their role in meaning-making. In order to spur communicators who likely understand their possibilities as negotiators of meaning, we have created a program that allows a certain degree of accountability while offering some flexibility in creating a route for negotiating cultural difference within the socially constructed spaces in which students work as technical communicators. Indeed, as Carl Lovitt notes, organizational cultures and professional discourse communities may shape communication in international contexts more significantly than national culture does (8). In his dynamic, process-based model of international professional communication, "international professional communication is constructed by the participants through dialogue, improvisation, and negotiation" (11), a view we think complements Kent's notion of how parology and triangulation as processes operate in any communication context. For Lovitt, international communication situations can "neither be described nor understood as the juxtaposition of two preexisting cultures; rather, the "culture" that defines this encounter is constructed by the participants during their interaction" (10). Through paralogic dialogue and improvisation, a third culture, whose dimensions cannot be anticipated in advance, is negotiated. These are theoretical positions that informed our conception of the certificate program.

THE "OTHER" SPACES OF TECHNICAL DISCOURSE AND ONLINE EDUCATION

As we undertake this curricular re-mapping process, we are also finding it useful to adopt a cultural studies perspective—specifically, a postcolonial one. When they began to administer writing instruction to future engineers, scientists, and students in other technical fields, English departments in effect "colonized" emerging technical discourses arising out of those disciplines. Several articles in the field have questioned the efficacy of the marriage between English and Technical Communication. In a 1994 article, Charles Sides concludes that successful relations between English and Technical Communication programs are rare, and that communications or media studies departments provide a more suitable marriage. And in "Can This Marriage Be Saved: Is an English Department a Good Home for Technical Communication?" Mary Sue MacNealy and Leon B. Heaton report on their results of an exploratory survey to assess the validity of both the anecdotal and documented evidence that "the relationship between teachers of technical writing and their English department colleagues is anything but blissful" (42). Of the sixty-six subjects who responded (39%), thirty were housed in English departments and thirty-six in other departments. The difference in these two groups' perceptions of departmental support for their programs proved statistically significant. Those faculty whose programs were housed outside English were significantly more satisfied with the level of support they received. Moreover, tenure and promotion problems for technical communication faculty in English departments have been significant enough for ATTW to publish a 180-page booklet on this issue.

Given these misgivings, we chose the postcolonial approach as a way to analyze (and perhaps successfully counsel toward a healthier relationship) this sometimes "odd couple(ing)" of English and Technical Communication. As Bernadette Longo has discussed in her book *Spurious Coin,* early technical communication textbook authors reinscribed in their texts the tension between liberal arts curricula and technical curricula in various ways. Crouch and Zetler's texts for technical writers clearly differentiated between technical writing and more general composition courses; they encouraged technical students to read more widely by referring them to an appended reading list of canonical literary works. The notion that science and engineering students' educations were "deficient for producing the type of well-rounded engineer who could understand human issues" (Longo 137) was prevalent among many academics, including science and engineering researchers and teachers themselves (135-137).

Interestingly, however, the tension between curricula played out quite differently in the work of Philip McDonald, author of the 1929 writing text

English and Science, who advocated that technical students would benefit more from reading the histories of civilization and science than works of great literature. Still, the intended outcome was the same—to bridge the gap between the arts and the sciences by implying and attempting to convince students of diverse technical specializations to accept the notion that they possessed a "common background of culture and humanism, which would…weld together the various groups of technical specialists…" (McDonald quoted in Longo 138). The focus on working to efface the differences between technical fields with an appeal to a common humanistic background is interesting we think. We see it as having much in common with politically motivated attempts to "unify" multiple subjectivities and "paper over" difference—difference which might more productively be acknowledged, legitimated, and negotiated. Indeed, the certificate program itself, with its diverse audience (both technically and culturally diverse) will not efface these differences, but rather allow for negotiation and articulation of difference among technical specializations and cultural backgrounds by providing a forum in which to apply, archive and observe the unpredictable and elusive strategies students employ in those negotiations—using the features of synchronous and asynchronous dialogue in the online chat and discussion board spaces of Blackboard.

In any process of colonization, forms of difference are articulated (Bhabha 66-68). With respect to scientific and technical communication programs, these forms of difference are often understood as:

- Differences in style
- Differences in the burden of information carried by text vs. visuals
- Differences in genre
- Differences in purpose
- Differences in audience (includes international now)

However, if we acknowledge, as Bhabha does, that these forms of difference have no "original" identity (as something called "technical communication") but are, instead, authorial choices that remain "polymorphous and perverse," that cut across a variety of writing situations variously identified with multiple programs and curricula within English studies, and that always perform a specific, strategic, and contingent calculation of their audience effect (Bhabha 67), then the boundaries drawn around scientific and technical communication programs can become more fluid and flexible, allowing for the establishment of stronger collaborations between and among programs such as the Rhetoric and Writing and Scientific and Technical Communication programs in our own department.

199

Such fluid and flexible spaces can become highly generative of innovative, hybrid instructional methods and artifacts like our online graduate certificate program.

Bhabha's description of the cultural 'beyond' that humanity found itself seeking at the twentieth century is illuminating for us as well. We see that as education moves online, the culture of the academy finds itself in a space/time transition analogous to that which Bhabha characterizes:

> We find ourselves in the moment of transit where space and time cross to produce complex figures of difference and identity, past and present, inside and outside, inclusion and exclusion. For there is a sense of disorientation, a disturbance of direction, in the 'beyond': an exploratory, restless movement caught so well in the French rendition of the words *au-dela*—here and there, on all sides . . . hither and thither, back and forth (1).

As members of the academy we find it useful to analyze the subject position of traditional, face-to-face education and its claim to "original" identity among other forms of learning. This analysis will help us "to think beyond narratives of originary and initial subjectivites and to focus on those moments or processes that are produced in the articulation of . . . differences" (1). We see our online graduate certificate program as an 'articulation of difference'—a collaborative, interstitial space capable of generating an alternative identity for our program. This identity cannot and should not be prescribed by us, but will be shaped by that restless, exploratory, back and forth movement Bhabha describes—a movement emerging from the disorientation and disturbance of direction experienced within our program, the English Department, and the university as we seek our own 'beyond' in the world of online education.

There are at least two interstices (Bhabha defines interstices as "the overlap and displacement of domains of difference") that our certificate program problematizes (2). One is the interstice that emerges from the border drawn between the English Department and the Department of Continuing Education. The other is the interstice generated by the border drawn around the programs of Scientific and Technical Communication and the English Department. We see our online graduate certificate program as a cultural hybridity—an artifact that is emerging in the moment of historical transformation that is online education. We want to exploit the "productive ambivalence" of online education as "Other"—to reveal the boundaries of traditional education's master narratives and to transgress these boundaries from the space of "otherness" that online education has been constructed as within the academic community.

CONCLUSION

At the same time we use the certificate to help us articulate a new identity for our program from an interstitial perspective, we also seek a new solidarity and community with other programs in the department, such as Rhetoric and Writing, from this interstitial perspective. On the one hand, we want to claim program agency from the in-between space of the overlapping domains of difference out of which our identity has been constructed by the colonial moves of English studies. On the other hand, we want to explore the imaginary of spatial distance—to live beyond the dual borderlands of traditional education/online education, and technical communication/English studies.

Toward these ends, we have successfully negotiated our institutional bureaucracy and received approval from the Graduate College to accept applications for admission to the certificate program. The program is advertised on the university website at both the Graduate College pages (link) and the Continuing Education pages (link). As such, these two links effectively serve as doorways into an as-yet-unnegotiated program space. So as we wait for applicants to materialize and triangulate these conceptual spaces, we continue to reinvent our institution through the perspective of a distant audience.

WORKS CITED

Bhabha, Homi K. *The Location of Culture*. New York: Routledge, 1994.

Duan, Ping and Weiping Gu. "The Development of Technical Communication in China's Universities." *Technical Communication* 52.4 (2005): 434-48.

Inquiry, Engagement, and Achievement: The Bowling Green State University Academic Plan. 2005. 22 February, 2006. <http://www.bgsu.edu/offices/provost/BGSUAcademicPlan1.PDF>

Kent, Thomas. *Paralogic Rhetoric: A Theory of Communicative Interaction*. Lewisburg, Bucknell University Press, 1993

Kynell, Theresa. "Technical Communication: Where Have We Been?" *Technical Communication Quarterly* 8.2 (1999): 143-151.

Longo, Bernadette. *Spurious Coin*. New York: SUNY Press, 2000.

Lovitt, Carl. "Rethinking the Role of Culture in International Professional Communication." Exploring the Rhetoric of International Communication. Ed. Carl R. Lovitt. New York: Baywood Publishing Company, Inc., 1999.

Readings, Bill. *The University in Ruins*. Cambridge: Harvard UP, 1996.

11 Introducing a Technical Writing Communication Course into a Canadian School of Engineering

Anne Parker

INTRODUCTION

Introducing technical communication into the curriculum of a Canadian engineering school has created its own set of challenges, particularly when some of the engineering professors continue to believe, as Mathes, Stevenson and Klaver suggested in 1979, that the subject is best taught by engineers. Doing so proved to be only modestly successful at my school. Yet, even without the push to use engineering faculty as my assistants, establishing one's authority as an expert in a non-engineering field can create a very real tension between the insider (the engineer) and the outsider (the technical communication instructor). As a female and as a non-engineer, I have occasionally felt like the "outsider." After all, a school of Engineering may well be the epicenter of what McIlwee and Robinson brand the "culture of Engineering," a culture that is both male-dominated and seemingly closed to the outsider.

The false perceptions of the engineering students only complicate the issue. On the one hand, many still perceive the subject as the study of "English," seemingly unaware that analyzing literature and writing essays about it is an activity quite different to writing engineering reports and giving technical presentations about technical problems and engineering designs. On the other hand, some students consider technical communication to be nothing more than grammar and composition, packaged though it may be in technical readings and exercises. Even some engineering professors also adhere to the latter view, and are surprised to discover that the field has grown to be such a rich and varied one (and one, incidentally, that demands the talents of a communication specialist).

Nevertheless, in spite of these misconceptions and challenges, I have found that, if an instructor can focus on the application of the technical communication field to the engineering profession, then many of these erroneous ideas can be dispelled. Indeed, over the years, the technical communication course at our school has met, if not anticipated, current trends within the engineering profession, exemplified most notably by the expectations of the national accreditation board. And this growing awareness of its relevance to the profession has resulted in the course's becoming more and more integral to the Faculty of Engineering at the same time as I have become less and less the outsider.

For example, in the 1970s, many potential employers simply wanted engineering graduates to be able to write more effectively, and the perceived absence of such a skill prompted complaints to the administrators of our engineering school. To address that need, our school then introduced a technical communication course in 1982, and students' writing skills noticeably improved. Later still, in the 1990s, the workplace had become more team-based, so the accreditation boards of both the U.S. and Canada urged engineering schools to introduce collaborative projects into the curriculum, partly because these projects helped to nurture the skills that were in high demand, such as interpersonal and project management skills. At the University of Manitoba, the technical communication course was already team-based, and thus served as a "prequel" to an emerging and significant trend – the inclusion of collaborative projects and instruction in teaming skills in the engineering curriculum.

Thus, to be successful, a technical communication specialist should be prepared to both adopt and adapt engineering practices. As this essay will discuss, the technical communication course offered in the engineering school at the University of Manitoba can serve as a case study to show how the tension between "insider" and "outsider" can be ameliorated and, more importantly, how the synergy between the practice of engineering and the communication of that practice can be effectively nurtured.

THE SYNERGY OF THE ENGINEERING FACULTY AND THE COMMUNICATION SPECIALIST: MEETING THE CHALLENGES AND ESTABLISHING AUTHORITY

When the technical communication course was first introduced as a compulsory component of the undergraduate program at my institution, we first thought that I would coordinate the delivery of the course as well as teach it. Given that at the time we had close to two hundred students per term, I couldn't do everything on my own, so we initially used engineering faculty as "assistants"; in other words, we did what Mathes, Stevenson and Klaver suggested we do. Such an arrangement was short-lived, to say the least. After one or two terms of teaching and marking the written assignments, most of these colleagues withdrew from the experiment, eager to return to their own courses and their own research. The technical communication course, in their view, was just too "demanding" and "time-consuming." In this sense, then, my colleagues were quite willing to acknowledge my expertise and "leave me to it." We then hired a part-time assistant for me, generally a graduate student in English, and, for a time, a graduate student in Civil Engineering. Interestingly enough, this latter arrangement worked surprisingly well, presumably because of his commitment to the course and to the principle of teaching engineering students the basic communication skills. After he graduated, we once again hired a series of assistants who had more of a humanities background.

However, in spite of the faculty's obvious willingness to leave me to teach the course, my place within the faculty hierarchy has been, at times, an ill-defined one, and one that occasionally even baffles my colleagues. For example, when I applied for tenure or promotion, they were at a loss as to how to evaluate me. They found they had to rely on the expertise of others in the technical communication field or in related fields. In Canada, that can at times be problematic because there are so few senior professors of technical communication. Most are instructors in two-year colleges or, if they do have a university appointment, they are usually junior members of the faculty – quite unlike my position at my school where I am now a fully tenured associate professor.

Another area where my position in the faculty has not always been clear is program and curriculum development. Over the years, even though the engineering faculty has frequently discussed the importance of building on what the technical communication course provides, there have been times when decisions about the curriculum have been made that did not include my input. Even today that happens, partly because these are professional engineers who are quite used to making decisions on their own; indeed, they expect to. They also see themselves as problem-solvers, and will therefore take what they consider as appropriate action to solve the problem. Once I remind them that I am the one with the "English" expertise, they will usually willingly accept my input and defer to my judgment in most matters of content and delivery.

In fact, establishing my authority and the place of technical communication within the engineering program has become easier over the years; now, there is much more of a cooperative effort between my engineering colleagues and me than there was at that time, although I have also had to work hard to promote both the field and myself. I have done this by joining engineering-related societies (like IEEE) and becoming an active member in them; by speaking to department meetings; by inviting colleagues into the class to observe what I do; by talking to them as often as I can about technical communication in general and the course in particular. They, in turn, keep me informed as to any developments within engineering that will impact what I do in the course.

So, all in all, the effort to introduce technical communication into the engineering undergraduate curriculum has been a worthwhile one. Furthermore, developing the technical communication course so that it accomplishes what both the profession and the faculty expect of it, while daunting at times and certainly time-consuming, has been an exciting challenge over the years. Now, the effort to integrate communication skills into the more senior level courses, including the graduation project, exemplifies the kind of synergy that is possible between communication specialists and engineering faculty (and, incidentally, the students, the end-users).

THE PROFESSIONAL CONTEXT: TECHNICAL COMMUNICATION AND THE PROBLEM-SOLVING MODEL

We can define engineering as the application of highly specialized, and technical, knowledge to a practical end that either remedies a problem or represents the "best" solution to a problem, usually within a set of defined constraints. We can then argue that engineers are essentially problem solvers. Following the Mathes and Stevenson model, enunciated in their book *Designing Technical Reports,* we can also say that most communication in the professional context of engineering comes about because of an engineering problem, a problem that someone has determined needs to be addressed (31). Thus, learning a problem-solving strategy – particularly one that helps to illustrate the connection between engineering design and the communication that must accompany it – will help students prepare for their professional lives in a way that a less practical approach might not achieve. While providing students with clear-cut steps to follow to accomplish their tasks, such a strategy must nonetheless be flexible enough to allow students to move freely between the steps, to pause and reflect, to test and explore, but without the kind of "lock-step" procedure that may stifle creativity and lead to a less satisfying conclusion (Winkler 119).

Some years ago, I began to develop such a problem-solving strategy after I realized that the processes used to describe both the writing practice, on the one hand, and the scientific and engineering practice, on the other, were remarkably similar, so much so that, in using the old scientific formula of "observe, test and solve," I could effectively talk to my engineering students about how they could proceed with their writing tasks. Indeed, these basic steps mirror those described by many other scholars who talked about the link between communication and problem solving (such as Barton and Barton; Dunkle and Pahnos; Flower; Maki and Schilling; Moran; Robinson; Souder; Tryzna and Batschelet;

and Winkler); those who discussed engineering design (such as Krick and Beakley et al); and those who studied engineering problem solving (such as Woods et al). These steps include: define the problem; brainstorm possible solutions to the problem; define the criteria to be used to assess any options offered as solutions to the problem; develop the prototype of a possible solution; test the prototype; and, finally, create the final product or document.

Eventually, I developed this connection into a more formal model that my students abbreviated to "C.A.T.S.," the acronym for Classify, Analyze, Test and Solve, as illustrated below ("Two Hats"; "Problem Solving"; "Implementing Collaborative Projects"; *Handbook*). This problem-solving model, stressing as it does both the methodology and the process, as Plants et al suggest, guides students as they work so that they are able to proceed fairly quickly and effectively. At the same time, the model is flexible, giving students the option to go back and forth between the steps or skip steps altogether. All in all, this model highlights the importance of problem solving to the *entire* engineering activity , and it is in this professional context that the model is so useful (Parker "Case Study Workshop" 40, Halstead & Martin 245).

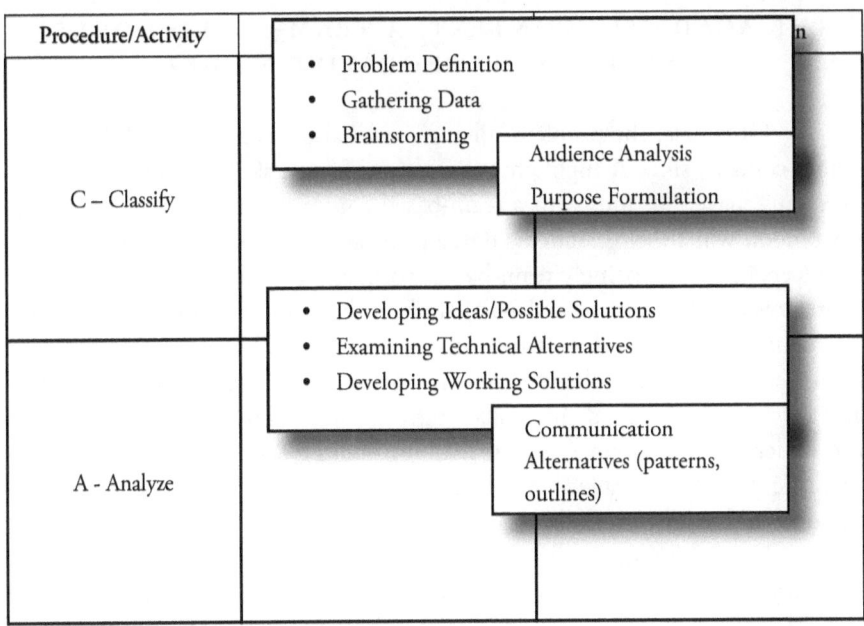

(more)

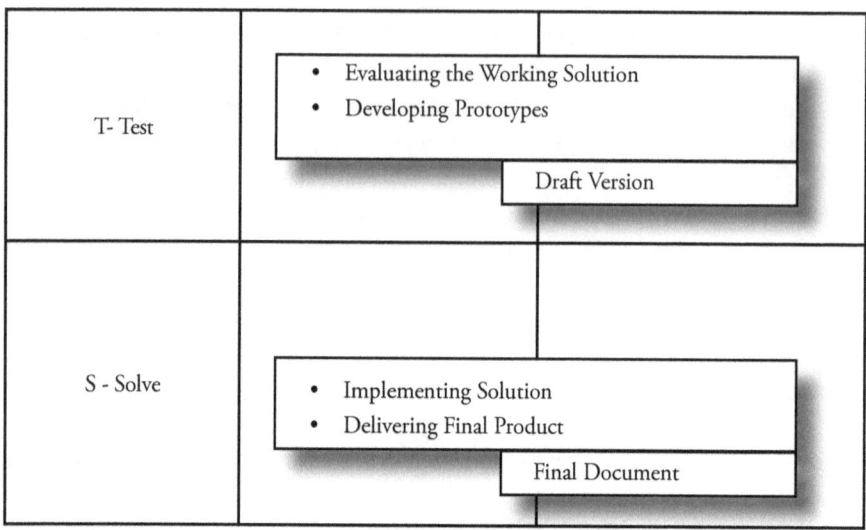

FIGURE 1: PROBLEM SOLVING IN ENGINEERING AND COMMUNICATION DESIGN

THE ACADEMIC CONTEXT: A TEAM-BASED COURSE AND THE COLLABORATIVE MODEL

Of course, the problem-solving model also has a place within the academic context, since it highlights at least one of the skills a prospective engineer must have. So, too, with a team-based course, such as the one offered at my school, which helps students develop the skills they will need as practicing engineers in an increasingly team-based workplace (Reimer 94, 99; Sageev and Romanowski 688; Vest et al 14-15). Indeed, both the Canadian Engineering Accreditation Board and its American counterpart, the Accreditation Board of Engineering and Technology, have come to recognize the need for such skills and have recently argued that collaborative projects should be integrated into the undergraduate engineering curriculum because such projects develop the requisite skills, the so-called "soft skills," such as project management, and interpersonal and teamwork skills.

In working on a team-based project, for example, students will work collaboratively through a "series of stages ranging from initial brainstorming to final report writing"; as they do so, "they become acquainted with such processes as participating in meetings, demonstrating leadership, and providing useful feedback to their colleagues"(Ingram and Parker "Influence of Gender" 7). Indeed, fairly recent work on the subject of collaboration, including Mary Lay's and my own, has supported the view that such projects help students to learn the values and protocols and language of their chosen profession. They do so by engaging in a process (collaboration) that ultimately leads to a product (a final report). If the process of communication instills the social element so critical to the success of the team's interactions, then the product of that communication represents the intellectual or learning outcome of that process. In this way, students become more familiar with their profession's discourse community, since they are researching an engineering topic and writing about it from an engineering point of view. Just as importantly, by writing and working as a team and by generating a product, students also become more "communicatively competent," more ready to assume their professional status (Bogdanowicz 1).

For these reasons, and also partly because I felt that such projects would encourage intelligent, but generally quieter, students to become more actively engaged in both the class and their own learning, I had already introduced collaborative projects some years earlier in the technical communication course offered at our school. Unlike their other engineering courses, the technical communication course enabled them to engage in a project where social processes (such as interpersonal and teaming skills) were as important as the intellectual ones. Along the way, at the same time as they learned more about an engineering topic, students would also be developing their oral and written communication skills.

However, most definitions (such as those of Allen, Blyler, Duin, and Flynn) tend to focus primarily on the social nature of collaboration – as I also did in some of my earlier work, defining collaboration as "a series of interactive activities that [are] social in nature" ("Influence of Gender" 9). The team's interactions help to provide the necessary "social knowledge" (Ingram and Parker "Gender and Modes" 34), gained as it is by "socially constructed" tasks ("Influence of Gender" 8). But to focus solely on the team and its individual members is to minimize the importance of what they produce, so this emphasis on its social nature should not ignore other important elements that will help to define and describe collaboration. While such factors as decision-making, responsibility and interaction are critical to an understanding of collaboration, so, too, is the purpose or the goal of collaborating; namely, to produce a document that, as a finished product, must "speak" for itself. Thus, the collaborative model that I will present here will include three essential elements - the project, the team and the collaborative process – all intertwined as illustrated below:

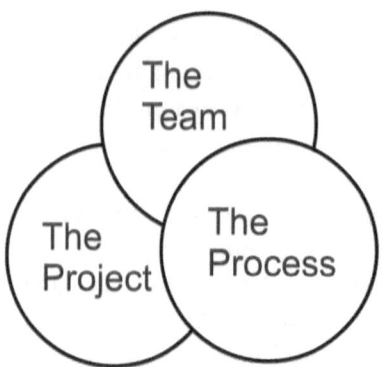

FIGURE 2: THE COLLABORATIVE ELEMENTS

The "project" will be a document or report; in other words, a finished product. It is also a product that someone else has requested and needs. Since, in technical communication, it is always reader-centered (and it is often a client who is the reader), this product will be the goal, a necessary outcome, of the collaboration. The students who interact so they can reach that goal are clearly the "team," and the "collaborative process" is the way they will reach that goal. Leadership theory likewise speaks of a variety of needs that relate to these elements. The project, for example, entails task needs or the jobs to be done; the social and emotional needs concern the team; and the procedural needs, such as how to accomplish the tasks, relate to the process of collaboration (Morgan 205-206). Together, these elements will describe what collaboration is within the academic context of an engineering classroom.

Nevertheless, introducing these collaborative projects into the technical communication course (and gradually changing the course into a team-based one) was not an easy task, and certainly not as straightforward as I had at first envisioned. For one thing, a classroom does not, and cannot, replicate the workplace, where things like group maintenance and team unity, the process, are less critical than producing what needs to be done, the product (Dannels 152 Freedman and Adam 402-418, Freedman and Artemeva 5). As well, the classroom imposes its own set of restrictions, from the physical space available for team meetings to the constant presence of an authority figure, namely, the professor. In the final analysis, perhaps all we can try to do, as Artemeva et al have suggested, is help our students transfer the skills we teach to the workplace ("From Page to Stage" 313).

But the greatest challenge confronting a professor who wants to introduce team projects into the classroom is evaluation. Initially, I believed that marking team reports would be less work than marking individual ones. In reality, though, what needs to be graded is not just the product itself, but also each individual team member who helped to create it. In fact, evaluation is perhaps the single most challenging aspect of collaborative projects, as I discuss at greater length in an earlier paper ("Evaluating Collaborative Projects"). And group projects certainly do not reduce your workload; rather, they increase it.

But a collaborative model and a team-based course do provide flexibility within the academic context. Unlike the more prescriptive lecture format, a team-based course demands that students have enough in-class time for group work. Once the professor has provided both the overview of the tasks at hand, such as reviewing each other's documents, as well as the framework needed to complete the work, students then have the chance to be actively involved in their own learning. Just as the workplace demands that teams be self-contained units, so, too, does the technical communication class. Students are expected to work on their own, resolve problems on their own, produce on their own. In other words, to be effective and to make that transfer of skills to the workplace possible, student teams should have roughly the same degree of autonomy as a workplace team would have.

Having said that, however, it is nonetheless important that the professor, unlike an employer, be available to intervene as needed. After all, these are still students who, unlike their professional counterparts, have no salary to compensate them for a poor group experience. Their grades depend on the smooth functioning of the team within the context of the classroom. Therefore, they shouldn't ever feel that they must "sink or swim"; rather, the professor is there to help them achieve their goals.

THE SYNERGY OF THE PROFESSIONAL AND THE ACADEMIC CONTEXTS IN THE TECHNICAL COMMUNICATION CLASS

The technical communication course that is offered at our school has certainly evolved over the years, but it has faced many challenges along the way, not the least of which is helping students to develop the "soft skills" they will need when they graduate. While I would argue that the problem-solving model described here provides students with a way of approaching their communication and design problems whether they are in the classroom or the workplace, the collaborative model reflects, rather than duplicates, the realities of the workplace. It nonetheless provides students with the kinds of skills they will need to succeed as professional engineers, such as project management, interpersonal skills and teamwork skills. Together, these models help to create a synergy between the professional and the academic contexts, a synergy that is as important as it is unique.

A representative series of tutorials that I have developed will illustrate this process. Organized according to the technical elements of the project, the communication elements and the team elements, the tutorials guide the students as they produce a document that both reflects the practice of the workplace and incorporates the attributes expected of the engineering graduate. For example, once students have either been assigned to a team according to their majors or have chosen their own team according to their common interests – and these team assignments occur early in the term, usually within the first two weeks of classes – we begin by detailing the two models and offering examples of how they work.

The early technical communication tutorials then focus on the team element and offer strategies to help students plan how they will proceed and how they will manage the team itself; for example, who will assume which leadership roles and how will they organize their meetings and their time. A subsequent tutorial encourages the teams to consider the collaborative process as a whole and, specifically, to discuss and begin to define such things as what their goals as a team will be and what standards of behavior they expect from each other. We build on this initial introduction to the process of collaboration later in the term, of course, but we try in these early tutorials to get students thinking about the whole "teaming" process and the kinds of skills they will need to develop if the group is to be a functional, productive team (and not merely a loose collection of individuals). Later in the course, other tutorials emphasize the various steps in the process of writing, revising and producing a document, including writing and revising strategies, document design, visual aids, and so on.

Other tutorials, meanwhile, have teams begin the work on their projects. Because the project must deal with an engineering topic, teams need to consider what technical issues they will have to consider. If, for example, they want to study traffic congestion on campus, they will have to decide how they will approach the issue; they might want to look at it from the perspective of traffic jams and line-ups or from the perspective of parking shortages. They will also need to determine how many cars do in fact create a problem and how you determine that number in the first place. From the discussion of the technical problem, these tutorials then talk about the need to evaluate any possible solutions, so teams must also develop criteria (such as cost or size or speed) by which to judge any options they are considering. As well, they need to define these as specifically as they can. About this time, too, another tutorial has the team looking at defining both audience and the document's purpose.

Thus, in preparing their document, students must first write a proposal suggesting they look at a particular engineering topic; then give periodic updates on their progress, including formal oral reports as well as informal briefings to the class; eventually deliver the completed written report; and, finally, orally present their findings to the class at the end of term. At the same time, as the Canadian Council of Engineers suggests, they have gained "a knowledge of the basic principles of project, human resource and time management" (3) through the series of tutorials, each of which emphasizes the different phases of the task while focusing on the technical, communication and team elements of the project.

CONCLUSION

In 1982, when the Faculty of Engineering at the University of Manitoba first introduced the technical communication course into its first-year program, few engineering schools in Canada had taken this bold step, although many schools in the U.S. had already developed technical communication courses for engineering students. But there was a distinct difference between the American precedent and what we were doing. Rather than being a member of another department altogether, like English, I was a member of the Faculty of Engineering. As well, rather than being offered as a service course, technical communication was a compulsory – and an integral - component in each student's program of study. In contrast, many technical communication courses in Canada, even now, are offered either by English departments or by writing centers that offer a variety of communication-related courses to a variety of disciplines.

Only recently, with the growth of academic programs dedicated to the study of technical and professional communication, do specialized departments with specialized instructors teach technical communication to future practitioners. But, again, this trend seems to be more pronounced in the U.S. than in Canada. One reason is the earlier development of such programs in the U.S. Conversely, in Canada, there are fewer programs offering only technical and professional communication, and most of these tend to be offered in the two-year colleges, although this may be slowly changing. So, all in all, there do seem to be some very real differences between the U.S. and Canada in terms of developing these professional writing programs.

Lilita Rodman, a leading Canadian scholar in the field of technical communication, addresses some of these differences when she suggests that what Canadian scholars focus on and what American scholars tend to emphasize in their research are not always the same. As Rodman notes, many of our scholars – scholars like C. Schryer, to name just one - have contributed greatly over the last while to current topics like the study of genre in technical communication. Besides these kinds of contributions, though, because Canada is a bilingual country, many of our Canadian scholars have also become increasingly interested in linguistics, a subject that seemingly has less interest for our U.S. counterparts (Rodman 13). Canadian scholars, then, do seem to have found their own particular niche in the field over the years.

Similarly, my position in the Faculty of Engineering at our school also seems to represent quite a unique niche – in Canada, at least - and the technical communication course that I have developed likewise holds a unique place in the development of technical and professional communication at our school since it is connected so closely to the undergraduate engineering curriculum. In essence, because I am so tied to the Faculty of Engineering itself, the course has come to be viewed as integral to engineering by students and staff alike. Additionally, over its twenty-year lifespan, this course has evolved into a smaller version of a technical communication program, one that is linked to both the Faculty of Engineering and the engineering profession. Indeed, as I have suggested here, it serves as a case study to illustrate the synergy that is possible between engineering and technical communication.

Initially offering instruction only in writing (and only to undergraduate students), now the course encompasses collaborative projects, project management, peer review, oral presentations, document design, textual illustrations and, recently, research methods. In the future, we hope to be able to offer a course in technical communication to our engineering graduate students; as an elective in a student's graduate engineering program, such a course will include topics related only to the academic side of engineering, such as thesis writing, preparing academic articles and oral defenses. Currently, we are also looking at ways to integrate technical communication into the graduation thesis and design project in a more formal way. These developments reflect trends in both technical communication and engineering (Ford & Riley 325-326).

Therefore, this paper has explored the academic and professional contexts for the study of technical communication in our school by looking at two primary topics: first, how a problem-solving model, as a way of approaching the communication task, adapts what is a common engineering practice to the teaching of technical communication and, secondly, how introducing collaborative projects into the technical communication classroom can be an effective way to prepare students for the demands of the workplace and the profession. Thus, within the professional context of engineering, the technical communication course can reflect the changing communication needs of the workplace and the engineering profession where team-based projects are increasingly the norm. Just as importantly, within the academic context, the course can reflect many of the developments in two seemingly disparate disciplines, technical communication and engineering.

WORKS CITED

Allen, N., Atkinson, D., Morgan, M., Moore, T., & Snow, C. "What Experienced Collaborators Say About Collaborative Writing." *Journal of Business and Technical Communication* 1:2 (1987): 70-90.

Artemeva, N. & Logie, S. "Introducing Engineering Students to Intellectual Teamwork: The Teaching and Practice of Peer Feedback in the Professional Communication Classroom." *Language and Learning Across the Disciplines* 6 (2002): 62-85.

Artemeva, N., Logie, S. & St-Martin, J. "From Page to Stage: How Theories of Genre and Situated Learning Help Introduce Engineering Students to Discipline-Specific Communication. *Technical Communication Quarterly* 8:3 (1999): 301-316.

Barton, B.F. & Barton, M.S. "The Case Method: Bridging the Gap Between Engineering Student and Professional." *Courses, Components, and Exercises in Technical Communication*. Ed. D. W. Stevenson. Urbana: National Council of Teachers of English, 1981. 22-33.

Barton, B.F. & Barton, M.S. "Toward Teaching a New Engineering Professionalism: A Joint Instructional Effort in Technical Design and Communication." *Technical and Professional Communication: Teaching in the Two-Year College, Four-Year College, Professional School*. Ed. T.M. Sawyer, Professional Communication Press, Inc., 1977. 119-128.

Beakley, G.C., Evans, D.L. & Keats, J.B. *Engineering: An Introduction to a Creative Profession*. 5th Edition. New York: Macmillan Publishing Company, 1986.

Blyler, N.R. "Theory and Curriculum: Reexamining the Curricular Separation of Business and Technical Communication." *Journal of Business and Technical Communication* 7:2 (1993): 218-245.

Bogdanowicz, M. "Communicative Competence: Business Savvy and the Technical Writingriter." *Technostyle* 6:1 (1987): 1-5.

Canadian Council of Professional Engineers (CCPE). *Report of the CCPE Accreditation Review Committee*. November, 1996.

Dannels, D. P. "Teaching and Learning Design Presentations in Engineering: Contradictions Between Academic and Workplace Activity Systems." *Journal of Business and Technical Communication* 17:2 (2003): 139-169.

Duin, A.H. "Computer-Supported Collaborative Writing: The Workplace and the Writing Classroom. *Journal of Business and Technical Communication* 5:2 (1991): 123-150.

Dunkle, S.B. & Pahnos, D.M. "Decision-Making and Problem-Solving: An Holistic Writing Assignment." *Courses, Components, and Exercises in Technical*

Communication. Ed. D.W. Stevenson, Urbana: National Council of Teachers of English, 1981. 205-209.

Flower, L. *Problem-Solving Strategies for Writing*. Harcourt Brace Jovanovich, Publishers, 1981.

Flynn, E., Savage, G., Pentik, M., Brown, C., & Watke, S. "Gender and Modes of Collaboration in a Chemical Engineering Design Course." *Journal of Business and Technical Communication* 5:4 (1991): 444-461.

Ford, J.D. & Riley, L.A. "Integrating Communication and Engineering Education: A Look at Curricula, Courses, and Support Systems." *Journal of Engineering Education* 92:4 (2003): 325-328.

Freedman, A. & Adam, C. "Learning to Write Professionally: 'Situated' Learning' and the Transition from University to Professional Discourse. *Journal of Business and Technical Communication* 10:4 (1996): 395-427.

Halstead, A. & Martin, L. "Learning Styles: A Tool for Selecting Students for Group Work." *International Journal of Electrical Engineering Education* 39:3 (2002): 245-252.

Ingram, S. & Parker, A. "Building an Effective Team: The Influence of Leadership Style on Modes of Collaboration." *Technostyle* 19:1 (2004): 82-108.

Ingram, S. & Parker, A. "The Influence of Gender on Collaborative Projects in an Engineering Classroom." *IEEE Transactions on Professional Communication* 45:1 (2002a): 7-20.

Ingram, S. & Parker, A. "Gender and Modes of Collaboration in an Engineering Classroom: A Profile of Two Women on Student Teams." *Journal of Business and Technical Communication* 16:1 (2002b): 33-68.

Krick, E. *An Introduction to Engineering: Methods, Concepts, and Issues*. New York: John Wiley, 1976.

Lay, M. "The Androgynous Collaborator: The Impact of Gender Studies on Collaboration." *New Visions of Collaborative Writing*. Ed. J. Forman, Portsmouth, NH: Boynton/Cook Publishers, 1992. 82-104.

Maki, P. & Schilling, C. *Writing in Organizations: Purposes, Strategies, and Processes*. McGraw-Hill Book Company, 1987.

Mathes, J.C., Stevenson, D.W., & Klaver, P. "Technical Writing: The Engineering Educator's Responsibility." *Engineering Education* 69:4 (1979): 331-334.

Mathes, J.C. & Stevenson, D.W. *Designing Technical reports: Writing for Audiences in Organizations*. Indianapolis: Bobbs-Merrill Educational Publishing, 1976.

McIlwee, J.S. & Robinson, J.G. *Women in Engineering: Gender, Power and Workplace Culture*. Albany: SUNY Press, 1992.

Moran, M.G. "A Problem-Solving Heuristic." *Technical Communication* 29:3 (1982): 38.

Morgan, M. "Women as Emergent Leaders in Student Collaborative Writing Groups." *Journal of Advanced Composition* 14:1 (1994): 203-219.

Parker, A. (with contributions from C. Strong & S. Ingram). *Handbook for Technical Communication.* Boston, MA: Pearson Custom Publishing, 2004.

Parker, A. "Evaluating Collaborative Projects and Evaluation Tools: Putting the Pieces of the Collaborative Puzzle Together." *Technostyle* 12:1(1995): 99-115.

Parker, A. "Implementing Collaborative Projects into the Technical Writing Classroom." *Technostyle* 10:2 (1992): 30-48.

Parker, A. "Problem Solving Applied to Teaching Technical Writing." *The Technical Writing Teacher* 17:2 (1990): 95-103.

Parker, A. "A Case Study Workshop and a Problem-Solving Approach to Technical Communication." *Technostyle* 8:1/2 (1989): 38-51.

Parker, A. "'Two Hats, One Head': The Problem-Solving Approach to Technical Writing." *Technostyle* 1 (1984): 7-16.

Plants, H.L., Dean, R.K., Sears, J.T., & Venable, W.S. "A Taxonomy of Problem-Solving Activities and Its Implications for Teaching." *The Teaching of Elementary Problem Solving in Engineering and Related Fields.* Ed. J.L. Lubkin, Washington, D.C.: ASEE Monograph, 1980. 21-34.

Reimer, M.J. "English and Communication Skills for the Global Engineer." *Global Journal of Engineering Education* 6:1 (2002): 91-100.

Robinson, P.A. *Fundamentals of Technical Writing.* Boston: Houghton Mifflin, 1985.

Robinson, P.A. "Technical Writing Workshops: An Alternative to Lectures." *Engineering Education* 73:4 (1983): 314-315.

Rodman, Lilita. "From Service Course to Discipline: The Evolution of Technical Communication (1960-2002)." *Bulletin,* Canadian Association of Teachers of Technical Writing (Spring 2003):13-14.

Sageev, P. & Romanowski, C.J. "A Message from Recent Engineering Graduates in the Workplace: Results of a Survey on Technical Communication Skills." *Journal of Engineering Education* 91:4 (2001): 685-693.

Schryer, C.F. "Genre and Power: A Chronotopic Analysis." *The Rhetoric and Ideology of Genre.* Ed. R. Coe, , Lingard, L. & Teslenko, T. Cresskill, NJ: Hampton Press, 2002. 73-102.

Souder, W. E. & Ziegler, R.W. "A Review of Creativity and Problem Solving Techniques." *Managing Professionals in Innovative Organizations: A Collection of Readings.* Ed. R. Katz, Harper Business, 1980. 267-279.

Trzyna, T. & Batschelet, M.W. *Writing for the Technical Professions.* Belmont, California: Wadsworth Publishing Company, 1987.

Vest, D., Long, M., Thomas, L. & Palmquist, M.E. "Relating Communication Training to Workplace Requirements: The Perspective of New Engineers. *IEEE Transactions on Professional Communication* 38:1 (1995): 11-17.

Winkler, V.M. "The Role of Models in Technical and Scientific Writing." *New Essays in Technical and Scientific Communication: Research, Theory, Practice.* Ed. P.V. Anderson, Brockmann, R.J. & Miller, C.R, Farmingdale, N.Y.: Baywood Publishing, Inc., 1983. 111-122.

Woods, D.R. "Problem Solving and Chemical Engineering, 1981." *Problem Solving* 79. Ed. J.T. Sears, Woods, D.R. & Noble, R.D., New York: American Institute of Chemical Engineers, 1983. 11-27.

Woods, D.R., Wright, J.D., Hoffman, T.W., Swartman, R.K. & Doig, I.D. "Teaching Problem Solving Skills." *Engineering Education* 66:3 (1975): 238-243.

12 English and Engineering, Pedagogy and Politics

Brian D. Ballentine

It is impossible, without giving offense to college authorities, to express one's self adequately on the English production of the engineering students...Most of them can be described only by the word "wretched." [1]

—English for Engineers

While some engineering schools have tried to manage their own writing programs, this chapter concerns itself with a professional and technical writing course created for junior-level engineering students at Case Western Reserve University, but housed, directed, and staffed from the English department. Although the course is a core requirement for all Case engineering majors, including aeronautical, biomedical, chemical, civil, computer, electrical, mechanical and software, it is administered from outside the school of engineering, automatically complicating staffing and curriculum.

These complications do not present insurmountable obstacles, however. To contextualize how we have established a workable system at Case Western, I begin with a discussion of the relationship between English departments and engineering schools in general. I then turn to the specifics of Case University's professional and technical writing course (English 398N). As I explain in more detail, one must be careful to present the course to students as a core engineering skill, one that has direct application to the engineering workplace. Given that audience awareness is key to success in technical writing, I address an effective and successful assignment on audience.[2] Finally, because our course is so large (over 350 students distributed into 18 sections each year!), I discuss the structure and goals of our graduate pedagogy seminar for PTW teachers, English 506. I end by discussing how this symbiosis of graduate and undergraduate courses functions without unnecessary complication, integrating well with the English department's rhetoric program yet distinguished from the required graduate composition pedagogy course.

ENGINEERING SCHOOLS AND ENGLISH DEPARTMENTS

As the professional field of engineering continues to grow, engineering schools frequently reassess core requirements and course curricula for their majors. Under these conditions, adding or even retaining courses focusing on "English production" becomes increasingly difficult. Engineering programs feel pressure from industry as well as competing institutions to produce graduates trained in the latest technology and engineering trends.[3] While an English department might consider an engineering school a unified monolith, the school's needs, wants, and ideas regarding a professional and technical writing program are actually fragmented. This fact should not be interpreted entirely as a fault, considering how varied fields of engineering attach varying degrees of importance to different types of communications.

As a result, engineering schools raise a number of concerns with English programs, the first of which is constructing a curriculum that can best meet the needs of a diverse group of engineering majors. In the past, English departments have attempted to teach similar courses by either using literature as a model for writing or teaching conventional rhetoric. Such approaches have been received unfavorably by engineering faculty or students. If anything, as Robert Connors' synoptic historicization of technical writing instruction made clear, these approaches helped create a "cultural split between English and engineering teachers."[4] That is, engineering students resist curriculum designed around English literature or technical writing scenarios where engineering is not the primary focus.

Although Connors' historical essay places the negative reception of technical writing in the past tense, engineering professors and students alike still refer to technical writing classes "disparagingly."[5] These remarks and the negative attitude towards such courses partly result from a curriculum that does not embrace the needs of a working and researching engineer. If the course is to succeed, the curriculum must be modeled around situations in business and industry where engineers will rely on communication skills to advance their work and careers. However, that criterion does not mean that English departments must compromise their own agendas for writing and communication. Case University's professional and technical writing course (English 398N) requires students and instructors to engage with both rhetorical elements of engineering discourse and the technical and scientific elements of an engineering project. As I will demonstrate, curricular flexibility and additional efforts on the part of the instructor to understand students' research and engineering interests are essential to integrating engineering topics and interests with professional and technical writing.

PROFESSIONAL AND TECHNICAL WRITING AT CASE: PROMOTING AN "OPEN" CURRICULUM IN ENGLISH 398N

Case Western Reserve University, a private research university located in Cleveland, Ohio, was formed in 1967 by the federation of Case Institute of Technology (founded in 1880 by philanthropist Leonard Case Jr.) and Western Reserve University (founded in 1826 in the area that was once the Connecticut Western Reserve). Ohio's largest independent research university, Case is most highly regarded for its medical school, ranked by *US News* at fifteen and twenty, respectively, in primary care and research, and for its engineering school, particularly the biomedical department, which consistently ranks in the top five among the nation's undergraduate biomedical engineering programs. For these reasons, Case receives the twelfth largest amount of federal research funding among private universities and spends nearly a million dollars *a day* on research.

Given the campus culture and environment, the Case English Department wisely made hands-on research the central concern when designing English 398N: Professional and Technical Communication for Engineers. This advanced writing course is structured around students identifying, proposing, researching, and presenting an engineering feasibility study spanning the entire semester. Research on the subject matter for the project needs to begin immediately. Students work in groups of their choice, preferably of three or four, and begin by completing a project outline form. The form helps break down the problem, the purpose and the audience for their proposed projects. Student conferences are recommended to assist groups in adjusting the proper scope of the project as well as addressing issues of scheduling and time management.

Flexibility and *choice* are central to English 398N. As mentioned, the growth of core requirements for engineering majors has impinged on students' ability to explore other interests. That is why our department invites students to use this PTW course as a chance to investigate a facet of engineering that is either not offered by their school or offered as an elective for which they may not have time. Flexibility is also encouraged regarding the makeup of the student groups. While some in the school of engineering may argue for corralling students into groups comprised of like majors, personal industry experience reveals that practicing engineers spend a large portion of their time collaborating with engineers outside of their own disciplines.[6] Indeed, for new products and solutions to function they often must integrate into other new or existing systems which automatically require cooperation from other engineers.

Pedagogically, students should be asked to identify and select their own research. Granted, instructors do guide and provide counsel for the projects. Nevertheless, instructors should not spoon-feed students prefabricated or "closed" case studies for these long-term projects.[7] Research by such scholars as Barbara Walvoord and Virginia Anderson suggests that allowing students to select their own areas of research can also significantly increase a student's intrinsic involvement in a course.[8] Martin Covington and Sonja Wiedenhaupt define *intrinsic motivation* as the "pursuit of intellectual inquiries which carry no immediate obligation to perform, nor any necessity for tangible payoffs except for the sake of satisfying one's curiosity or for the productive exercise of the mind."[9]

Of course, instructors must continue to attach grades and performance reviews (*extrinsic* motivators) to such assignments. However, giving students an opportunity to pursue areas of interest for which they might not otherwise have time can create a dynamic environment for the course and for class projects. Ken Bain's article in the *Chronicle of Higher Education*, "What Makes Great Teachers Great?" identifies the creation of a "natural critical learning environment" as the foundation for a successful course. Bain explains:

> "Natural" because what matters most is for students to tackle questions and tasks that they *naturally* find of interest, make decisions, defend their choices, sometimes come up short, receive feedback on their efforts, and try again. "Critical" because by thinking critically, students learn to reason from evidence and to examine the quality of their reasoning, to make improvements while thinking, and to ask probing and insightful questions.[10]

The challenge for English 398N instructors is that engineers view different project tasks as intrinsically or naturally more interesting than others. As Dorothy Winsor documents in her landmark ethnography, *Writing like an Engineer: A Rhetorical Education,* young engineers tend to devalue documentation, reporting, and presenting, the very communication tasks which technical writing instructors believe are essential to invention, knowledge production, and productive engineering. Instead, the *invention and creation of material artifacts*—glorified by engineering since the days of Vitruvius—captures our students' attention. Even so, successful invention, discovery, and problem solving require communication skills in the engineering workplace, to say nothing of the public realm. Vitruvius, after all, never would have become the father of Western engineering if he could not write a courtly cover letter to Augustus, the father of all clients. When students grasp this political and rhetorical lesson, English 398N becomes central to their education and future careers.

A well-designed PTW curriculum, therefore, not only will encourage students to respect all components of an engineering project but demonstrate to students that persuasive communications are not external and extraneous to the engineering process, but rather internal and inherent to their field. To accomplish this crucial goal, instructors must use course assignments to orient students towards their prospective readers. Accordingly, the next section details all of the assignments in the course's semester-long sequence: such print texts as the *project topic form, client letter, proposal, progress report, feasibility study,* and *feasibility study presentation* and such electronic texts as a *web site* and a *web site presentation.*

COURSE ASSIGNMENT SEQUENCE: BRINGING AUDIENCE AND ENGINEERING TOGETHER

By creating a series of persuasive documents throughout the semester, English 398N students learn to develop solutions to the unique challenges and circumstances they encounter as their research progresses. The student research and the semester-long project strive to answer one question: *Is this engineering project feasible?* At the beginning of the semester, the answer to this question is indeterminate. By the end of the assignment sequence, students must present their results, their discoveries, their recommendations, or in other words their answer, to the rest of the class. Each student group must convince the class that their solution and recommendation on how to respond to a particular engineering problem are not only viable but optimal.

The groups' first assignment requires them to begin completing *a project topic form.* The form contains five short categories designed to help students identify: 1) the engineering problem, 2) the purpose of the research, 3) the specific audience, 4) the desired change within that audience, 5) available resources for their proposed projects. As students begin to detail their problem, they must discuss how their engineering studies relate to their proposed project. This cornerstone assignment ensures students witness the centrality of engineering to writing and of writing to engineering. For the best results, instructors should schedule student conferences in order to assist groups in adjusting the proper scope of the project as well as addressing issues of scheduling and time management. While the project topic form asks students to begin analyzing their audience, the actual engineering project retains prominence.

However, as student groups begin completing their project topic forms, instructors need to reinforce the importance of not just investigating a particular engineering problem but also identifying the audience receiving the communications. Focusing on both audience and engineering addresses what I consider a critical concern: Too many textbooks tend to conceive of audience and audience interaction as external to the engineering enterprise. Thus, audience in all its specificity is never adequately treated, or is treated only impressionistically. To compensate for this deficiency, student groups refer to their primary audience as the "client," a common practice in business and engineering.

Since the group projects run the duration of the semester, students are required to nurture relationships with their determined client from the start. Appropriately, before actually drafting their *research proposals,* student groups must introduce themselves to their clients. The group collaborates on a letter in which the students introduce themselves, give an overview of the identified problem, let the client know about the coming proposal, and begin to establish credibility. Although students are given the option of researching and responding to a formal request for proposal or RFP, the student projects are largely unsolicited proposals. In this course, the *client introduction letter* simulates the real-world scenario of drafting an unsolicited proposal. Rife with uncertainty, unsolicited proposals are often much harder to construct than proposals tailored to a specific RFP.

The *project topic form,* the *client letter,* and the actual *proposal* are the first three assignments in the course's interlocking assignment sequence. At the beginning of the course, the instructor must explain each assignment's individual purpose and how the assignments dovetail into each other. The client letter, for example, prepares the audience for the reception of the proposal, while the goal of the proposal itself is to gain the client's permission to proceed with the proposed *engineering feasibility study.* While this sequence does not mirror industry one-hundred percent of the time, it is common for companies to propose studies that report on the feasibility of an engineering project before investing more of their resources.

Generally, the proposal begins with background information and components of the current situation the student group proposes to investigate. The groups elucidate the engineering problem and state their objectives surrounding that problem. The degrees to which all of the components of a standard proposal, such as criteria, method, solution, schedule, cost, conclusions, and recommendations, come into play vary from project to project. Instructors should note that the use of forms, like a standard proposal which may seem "natural" as a basis for instruction in the classroom, are only effective if they have a relevant engineering purpose and situation attached to them.[11] That is, leading instruction with forms without exigency will not persuade students of the importance of communications in engineering.

Charles Bazerman states the issue succinctly, if sternly: "As teachers, if we provide our students with only the formal trappings of the genres they need to work in, we offer them nothing more than unreflecting slavery to current practice and no means to ride the change that inevitably will come in the forty to fifty years they will practice their professions."[12] To overcome these "trappings," researched engineering projects must teach students how to apply genres to an authentic engineering research project. Again, student conferences are recommended so instructors may manage the projects as they grow. For the purposes of this class and this structure, all proposals end with the request that the audience authorize the group to move forward with a feasibility report. That is, with successful proposals the groups have effectively persuaded their audiences that the identified issue is serious enough or potentially beneficial enough to justify the cost of doing the research for the feasibility study.

After students receive permission to move on to the feasibility report the interlocking assignment sequence contains additional work for the students.[13] Just as in industry, clients want updates on the engineers' work. Student groups are responsible for a formal progress report written for their defined audience. Clients who have invested resources in a project desire ready access to reporting. To that end, student groups are asked to develop a web site that supports their studies. In terms of communicative goals, the site is an efficient means of providing up to date progress information. In industry, many engineering firms use web sites to manage tasks and schedules for their projects. Students prepare and deliver short presentations on the design of their web sites in order to prepare for their upcoming final presentations.

The last stage of the assignment sequence is for each group to deliver a *formal presentation* to the rest of the class reporting on the results found in their feasibility report. Each group must make a recommendation and defend their findings in a question and answer session. Each of these stages presents instructors with the opportunity to teach systematically all of the standard forms for a professional and technical writing course including proposals, progress reports, feasibility studies, and formal presentations. The curriculum for this course avoids promoting the mere "trappings" mentioned by Bazerman by allowing student groups to begin and end with an engineering project containing goals and objectives that are their own.

The assignment sequence for professional and technical writing provides evidence for the "open" versus "closed" approach to the course. In this open model, students identify and investigate a real engineering project and audience. In a closed model with a fabricated audience, however, "if students want to know more about these fictional readers' motives, values, or attitudes, they find that these important issues are not available. Or, worse yet, they discover that the teacher is making up answers to these important audience-analysis issues off the cuff."[14] Instead of the instructor fabricating audience characteristics the students are charged with researching their real audience just as they would in a professional engineering situation. The open model thus reduces the chance of "pseudotransactional" writing or "writing that is patently designed by a student to meet teacher expectations rather than perform the 'real' function the teacher has suggested."[15] In the professional and technical writing course the "real function" is to investigate the feasibility of an engineering project.

SAMPLE STUDENT PROJECT: A FEASIBILITY STUDY ON LINUX

The above section provided an outline of English 398N's major assignment structure; but to better demonstrate the curriculum for the professional and technical writing course, I have included a sample project and traced its steps. This example is an original student project and is detailed here with permission of the student group.

The Linux Project

In this project, a work group begins with premise of proposing a study to overhaul the university's server platforms because as young software and computer engineers they are dissatisfied with Windows NT. According to these students, other equally robust Linux-based applications could meet the university's needs. Ambitiously, the group wants to explore the possibility of developing their own brand of Linux for the campus. They cite security issues and potential cost benefits as primary arguments for their case and indicate their audience would be the chief information officer and his or her staff. The project form is brief but engages student groups with their engineering ideas.

Project Topic Form

The engineering "problem" you intend to investigate:
   The feasibility of replacing the Windows NT servers that support the university's network infrastructure with a student developed system based on the Linux operating system.

Purpose of and need for this project:

Conservative estimates in regards to money lost due to security issues with Windows software are calculated to be in hundreds of millions of dollars. Instead of relying on and waiting for Microsoft to develop and release patches for their software the university could actively develop its own repairs when problems arise. Code for Linux is distributed under the terms of a General Public License (GPL) that states the code is free as long as any improvements or alterations that are made to the code are not hidden from others. Linux is part of a unique on-line phenomenon known as "open source" development where programmers from around the world share their work to improve applications such as Linux. Microsoft, on the other hand, intentionally obfuscates its code from users and developers.

Description of target audience(s):
University CIO, department of computer science and engineering, engineering students, and non-engineering students.

Desired changes in target audience(s):
That the audience will recognize the opportunities to not only save money and reduce network "down time" but also provide students with a unique learning opportunity. The audience will attain a clear understanding of the technical and economic feasibility of this plan.

Available resources to support this project (internet, library, personal, etc.):
Library, university network administrators, software engineering professors, Linux development web sites.

Student Conferences

The course instructor and the student group meet to confer on their project. During the meeting, the group specifies that they want to replace the server platform with Linux not only to make it more secure but to integrate it into the global open source software movement. The group will need to define this movement to all members of their audience and explain why it would benefit the school to become a part of it. The instructor suggests that the option of developing their own brand of Linux could be difficult to maintain long term. The team therefore decides to explore existing Linux packages, such as Red Hat and SuSE, as an option. They also expand their identified audience beyond members of the IT department to include high-level administrators who would also be involved in the decision-making process.

Client Letter

The letter introduces group members as computer and software engineering students, who have recognized a potential security problem in the current system. The group believes that a Linux implementation may save the school money as well as bring the engineering school recognition for the innovative project. They inform their audience members about the forthcoming proposal, when to expect it, and that they hope to gain authorization to complete a feasibility report.

Proposal

The group proposal discusses the background information on the university's use of NT and some of the known security issues surrounding that technology. Its objectives include eliminating security breaches, cutting down on maintenance and down-time, and increasing network compatibility. For their proposed feasibility study to succeed, the students will have to examine statistics on Linux security, investigate other institutions or businesses that have implemented it, determine initial costs, long-term maintenance costs, training needs, and time to implement to name a few. The work done towards developing the proposal helps groups identify their criteria and objectives for success with the final projects. In short, the group begins to get a clear sense of what they will have to uncover for their study to succeed.

Progress Report

After the students turn in their proposal and the instructor authorizes them to proceed with their feasibility report, communication with the client becomes even more critical. The progress report is assigned in an effort to demonstrate that business and engineering groups are held accountable for their work. In this particular case, the audience for the report is understood to be the university CIO but students must bear in mind that anyone on staff could potentially be a reader. Again, since student groups are preparing engineering planning documents for each assignment they should be discovering all of these potential audience members. Instructors warn that a manager can ask for a "progress report" at any time and while people are generally pleased to hear about past success they are more interested in whether or not the group will meet upcoming deadlines and if the project will finish on schedule.

Website

A great deal of communication occurs online, and many engineering projects are managed via the internet or corporate intranets. While the web portion of the course is in place for this reason, it also requires students to conceptualize how to organize and structure their data in a digital environment. The group members will need to ask themselves questions unique to a digital environment such as, "What is the first thing the audience for this project will want to see if they come to the site for a quick update?", "Where is all of the empirical data going to be placed?", "How does the audience contact the group and who is in charge of what parts?" At this stage, it may benefit the student groups to dissect other web sites, particularly those related to similar engineering issues. Students will want to discuss what makes some sites succeed and others fail in terms of navigation, organization, and information availability. Engineers are often managing large data-sets and this assignment helps introduce that task.

Website Presentation

The web assignment is an excellent opportunity to discuss the power of communication as it relates to the specific group projects as well as to give student groups the opportunity to keep abreast of their peers' research. Instructors may find it advantageous to ask the groups to give informal presentations on their web sites so they may explain their communication and hierarchy decisions to the rest of the class. This is an effective way to generate dialogue between the student groups. This presentation exercise introduces the student groups to presenting as a team, which many of them have not had to do. In the engineering workplace, teams often present as a group and not individually. Also, instructors should note that with the web presentation coming first, the final presentations on the feasibility study recommendations generally excel due to the additional practice.

Feasibility Study

Finally, the feasibility report is due. In the Linux case, instructors can expect to see a detailed cost breakdown in terms of servers, initial installation fees, software fees, maintenance, additional personnel, training, and total cost of ownership models. There also should be a convincing amount of data on security benefits realized by other institutions or corporations, unbiased testing centers and user testimony. There should be a concise timeline for installation and implementation. Naturally, there will be a recommendation on whether or not to develop a brand of Linux, buy and implement an existing Linux package, or leave the NT system unchanged. Instructors advise the students that in business feasibility studies are researched and written by engineers so companies do not make costly investment mistakes. That is, despite all of their hard work and fondness for Linux, student groups may conclude that the best solution is to stay with the present system.

Feasibility Study Presentation

All of the group's findings are reported to the class in the final presentation. Instructors may wish to create a setting that is more formal than casual and place a good deal of importance on the groups persuading their peers that their data and their recommendations are sound. Peer groups have the opportunity to challenge the presenting groups and their recommendations in a question and answer session.

INTEGRATION WITHIN THE RHETORIC CONCENTRATION AND THE TRAINING OF PTW INSTRUCTORS

Despite its careful calibration, the curriculum design for English 398N would have proven ineffective, if the Case English department had not implemented an effective instructor training and credentialing system. The question we faced is how can we train fledgling rhetoric scholars, well-versed in Aristotelian concepts and the uses of *logos, ethos,* and *pathos,* to teach engineering students communication skills? The problem is complicated by considering who is most likely to teach such a professional and technical writing course. With even smaller-sized engineering programs placing heavy teaching demands on English departments, that demand is often met by graduate students (at Case the number of students enrolled in engineering necessitate offering eighteen sections of the course a year with approximately twenty students per section). An advanced professional and technical writing course presents young instructors, most of whom have only limited experience teaching composition courses, with many new challenges, including a diverse engineering audience as well as advanced software and technology.

According to Connors, as technical writing "grew up" in the second half of the twentieth century, the "age-old battle raged on between those who wished to teach technical students to write and those who wished to teach them to read and appreciate great literature."[16] Despite the growth and acceptance of professional and technical writing programs coexisting with English departments, a divide can be felt between those who teach writing and those who teach literature. This national trend is evinced primarily by the low number of senior faculty involved with or interested in teaching professional and technical writing in traditional English departments. For scholars pursuing careers in professional and technical writing, the struggle for acceptance and legitimacy within an English department remains challenging. But, as members of PTW programs are discovering, there are opportunities to effectively integrate with and contribute to English departments without being dismissed as "second rate."[17]

For example, Case's graduate technical writing pedagogy seminar is not a freestanding anomaly without ties to our department's other research foci. In fact, for a PTW pedagogy seminar and a PTW curriculum to secure reception as legitimate scholarly activities worthy of *any* English department, both must be understood in terms of that department's larger offerings. To illustrate, the Case English department offers a concentration in "Writing History and Theory," referred to as the WHiT program. This program resembles other rhetoric programs which have been gaining momentum in academia; but because it also examines the practice if writing as *historically, culturally,* and *technologically* situated, it can more easily accommodate a professional and technical writing pedagogy course.

PhD students in the WHiT program study rhetorical theory and history, the history of writing and publishing practices, authorship, linguistics and semiotics, and digital communication theory. The program examines relationships between textual elements such as word-image interface, lexical and grammatical choices, document design, and global and rhetorical issues, such as text production and circulation, copyright, audience, ethics, and rhetorical effect.[18] Equally important, especially for PTW purposes, the program allows graduate students to develop an in-depth understanding of the way that writing functions not only in cultures and society but in individual disciplines, organizations, and institutions.

Graduate students interested in WHiT, must enroll in a course on rhetorical theory, which serves as an overview and a backbone for the program. Beyond this core requirement, the program itself is comprised of three general areas designed to provide students with the necessary theoretical and historical foundation on the study of writing: history of writing, digital writing, and linguistics and semiotics. As I will demonstrate, the WHiT program is an excellent opportunity not only to prepare graduate students for the specialized classroom of professional and technical writing but also to promote the examination of engineering as a rhetorical practice and analyze the unique discourse of the engineering discipline.

Most English departments, however, offer "standard" pedagogy courses, designed to prepare graduate students to teach freshman composition. This practice can raise concerns from faculty invested strictly in literature or composition studies about the need for a second, separate seminar for professional and technical writing. Indeed, whether in the WHiT program or the literature-based concentration, all graduate students at Case are required to enroll in the pedagogy course titled "Rhetoric and the Teaching of Writing," designed to ready graduate students for the composition classroom. Most standard pedagogy classes focus on graduate students gaining an understanding of major themes in composition theory in order to develop a set of coherent, historicized pedagogical practices. Typically, the primary goals will include developing an understanding of the major trends in composition scholarship and pedagogy, and to explore a variety of pedagogical strategies for writing classes, including assignment sequencing, assessment techniques, and student conferencing.

At Case, this course challenges graduate students to develop a research project proposal of their own that demonstrates engagement with current issues in composition and rhetoric as well as constructing a syllabus and assignment sequence to be used in a future writing course. Unquestionably, these goals are so broad that an English department may raise legitimate concerns regarding redundant curriculum in a second pedagogy course, especially when graduate students are calling for a larger selection of course offerings. Consequently, a professional and technical writing pedagogy course should be distinctive and separate from standard pedagogy offerings as well as advance the WHiT agenda.

English 506, Teaching Technical and Professional Communication, sets itself up first as a WHiT seminar with a strong scholarly agenda and second as a practical guide to instructing professional and technical writing. This graduate course strives to align itself with the general mission of the WHiT program, promoting topics that include studies in rhetoric of science and technology; history of professional and technical writing; critical approaches to technology; ethics and law (e.g. copyright and intellectual property); collaboration and management of writing projects; document design theory (print and electronic); theories on digital reading and writing; engineering and science concentrations; and, finally, practical matters of curriculum design, assignments, writing evaluation and course management.

In addition, a portion of each seminar is set aside to address practical pedagogical issues. Among the requirements for the course, graduate students must observe at least two professional and technical writing classes. Afterwards, students reflect on those observations in the seminar and ask questions of the instructor for clarification on class proceedings. Graduate students are responsible for reviewing a portfolio of engineering writing and evaluating the work with their peers from the pedagogy course.

General pedagogy courses such as Case's challenge graduate students with scholarly activities such as researching and proposing new approaches to composition theory or historicizing the growth of composition and rhetoric courses. Similarly, the professional and technical writing pedagogy course must demonstrate a unique scholarly agenda. The course taps into the rich field of the rhetoric of science by expanding the term more broadly to the rhetoric of science, technology, and engineering. In his introduction to *Landmark Essays on the Rhetoric of Science,* Randy Allen Harris defines the rhetoric of science as "the study of how scientists persuade and dissuade each other and the rest of us about nature, – the study of how scientists argue in the making of knowledge."[19]

As an established field, therefore, the rhetoric of science provides a starting point for analyzing and discussing technical writing. Graduate students will be given the opportunity to explore the similarities and differences between scientific and engineering rhetoric and discourse. Discoveries and inventions, Harris maintains, need to be analyzed and argued not only among the scientific (and engineering) community but amongst the "rest of us" too.

Besides such notables as Harris describes, numerous other works augment the exploration and development of the curriculum for Case's PTW pedagogy course. Prominent figures include Alan Gross, Jeanne Fahnestock, Dilip Gaonkar, Carolyn Miller, and many others whose research provide avenues into the study of what degree persuasion plays a part in science. Charles Bazerman's *The Languages of Edison's Light* is an excellent means of examining the role rhetoric plays in engineering and invention. Bazerman artfully excerpts pages from Thomas Edison's journals, patent applications, and personal letters which all "reveal the rhetorical activity of the discourse" surrounding Edison's discoveries.[20]

Such texts are necessary. Our graduate students, who have the opportunity to teach engineers, report struggling with convincing the class of the value of this "rhetorical activity." English composition instructors as well as the students in their classrooms are generally more comfortable with the notion that a "right" answer is the answer which is best argued. The transition to a professional and technical writing classroom comprised of engineering students can challenge this belief. Engineers are more likely to search for a concrete and proven "right" answer or the equivalent of some "transcendent absolute truth" in the spirit of Plato.[21] In short, the "brute facts" are valued.[22] In contrast, Edison's patent applications took advantage of the patent review system in that Edison argued "based on the novelty of a conception rather than on its proven viability, usefulness, or market value."[23]

Regardless of these industry realities, engineering students often have less patience for rhetorical practices that debate what is possible, probable, or even most likely to be true. Consequently, while the study of rhetoric forms an effective foundation for preparing graduate students to teach the course, overtly promoting rhetoric to a classroom of engineering students can meet with a cold reception. Graduate students learn that their studies of rhetorical theory, especially as it relates to scientific activity, provides valuable insight to a new community of engineering students but that explicit rhetorical terminology cannot serve as the structure for the course. Indeed, most of their rhetorical strategies focus on establishing *ethos* for themselves as authoritative instructors. My past experience as a senior software engineer at Marconi Medical Systems, a medical imaging company and a subsidiary of Philips Electronics, probably carries more weight with my students than my PhD in English.

PTW instructors can overcome student resistance and skepticism, however, by keeping their classes "user-friendly." English 398N's combination of the project topic form and assignment sequence is a tested method for maintaining a "project-centered" focus, one which increases intrinsic motivation for the course. Indeed, it is the "open" curriculum of the course that assists PhD students specializing in rhetoric with succeeding in their teaching.

Despite significant progress over the last several decades, arguments are still being made that "the technical communication course should be taken out of the hands of English teachers."[24] Given the persistence of this attitude, instructors must be equipped to engage with engineering and their engineering students' projects. Consequently, pedagogy courses that specifically address the needs of individuals preparing to teach professional and technical writing courses are essential to graduate students.

Such courses are also a welcome addition to rhetoric programs such as WHiT so that the programs may succeed in their missions to prepare graduate students for the academic job market, which increasingly favors candidates possessing the ability to teach in different areas including composition, linguistics, and technical writing. Case University's two-part strategy of redesigning the professional and technical writing course's curriculum (foreground engineering situations, research, and interests in PTW courses and initiate a new pedagogy course for English graduate students) is the best method for enhancing and sustaining the complex relationship between English departments and schools of engineering.

This improved collaboration would benefit not only the academy but the nation. As Hurricane Katrina graphically showed, our country's infrastructure has become disgracefully derelict. More than ever, we need civic-minded engineers who can make their case to government and industry, voters and consumers. Effective professional and technical writing instruction, therefore, has become a necessity, not a luxury. As Samuel C. Florman observed a decade ago:

> By creating the engineers of the future, educators can transform the world in meaningful ways. Yet engineering education cannot flourish in the absence of popular regard and government support. We have something like a Catch-22 here. Appropriate education is needed to further a renaissance in engineering, but a renaissance in engineering is needed to inspire steps toward appropriate education. Someone must break this paralyzing cycle.[25]

Public works require public words. As the ancient Romans realized, the orator and the engineer are alike. Both deal with *res publica*—the orator by constructing arguments, the engineer by arguing for construction. To prevent our own republic from crumbling, English departments must build bridges between rhetoric and engineering. The best way, as outlined here, is to offer an open and flexible professional and technical writing curriculum.

NOTES

[1] English for Engineers. (1915, June 19). *Engineering Record* p. 763. quoted in Connors, Robert J. "Landmark Essay: The Rise of Technical Writing Instruction in America." *Three Keys to the Past*. Eds. Teresa C. Kynell and Michael G. Moran. Samford, CT: Ablex Publishing, 1999. 175.

[2] Here I am referring to the three best-selling textbooks frequently used for the professional and technical writing classroom by Paul Anderson, John Lannon, and Mike Markel. See:

    Anderson, Paul V. *Technical Communication: A Reader Centered Approach*. (5th Ed.) Boston: Heinle, 2003.

    Lannon, John M. *Technical Communication*. (9th Ed.) New York: Longman, 2003.

    Markel, Mike. *Technical Communication*. (7th Ed.) Boston: Bedford/St. Martin's, 2003.

[3] Computer science is an excellent example of an engineering field struggling to keep pace with growing technology. Traditionally, C++ serves as the core computer language taught in introductory courses and it is used or referenced throughout the remainder of a student's education. Over the last five years, schools have been abandoning C++ in favor of Java. However, even more recently schools are experimenting with Microsoft's newer language C# (pronounced C sharp). All of these languages are object-oriented in nature but each possesses

unique attributes. Regardless of which language a school selects, companies will continue developing applications with all three and engineering schools will feel the pressure to keep pace. Consequently, these demands cause reevaluations of a school's course offerings and often force courses out of a system.

[4] Connors, Robert J. "Landmark Essay: The Rise of Technical Writing Instruction in America." *Three Keys to the Past.* Eds. Teresa C. Kynell and Michael G. Moran. Samford, CT: Ablex Publishing, 1999. 176-7.

[5] Ibid. 178.

[6] I can testify to this fact personally. For three years I worked as a senior software engineer for a major medical company. Companies such as my employer hired engineers from most of the major engineering disciplines. In order for our software applications and other company initiatives to integrate with the rest of our products and services, our software engineering team needed to collaborate and communicate with other engineering disciplines including electrical and biomedical engineering. Also, it should be noted that guideline "d" on page two from The Accreditation Board of Engineering and Technology criteria states students must attain: "an ability to function on multi-disciplinary teams."

[7] In their article, "Genre, Rhetorical Interpretation, and the Open Case: Teaching the Analytical Report," Sheehan and Flood assert that, "To situate their students, technical writing teachers have typically turned to closed case assignments" (21). Their research instead advocates "the use of open cases in which students use the analytical report genre to interpret and study technical issues in a workplace where they are already situated – the university campus." In, *IEEE Transactions on Professional Communication,* Vol. 42, No. 1, 1999, 21.

[8] Anderson, Virginia and Barbara Walvoord. *Effective Grading: A Tool for Learning and Assessment.* San Francisco: Jossey-Bass, 1998.

[9] Covington, M. V., and Wiedenhaupt, S. "Turning Work into Play: The Nature and Nurturing of Intrinsic Task Engagement." R. Perry & J. C. Smart (Eds.), *Effective Teaching in Higher Education: Research and Practices.* New York: Agathon Press, 1997.

[10] Bain, Ken. "What Makes Great Teachers Great?" *The Chronicle of Higher Education.* April 9, 2004. vol. 50, issue 31 pp. B7.

[11] Connors, Robert J. "Landmark Essay: The Rise of Technical Writing Instruction in America." *Three Keys to the Past.* Eds. Teresa C. Kynell and Michael G. Moran. Samford, CT: Ablex Publishing, 1999. 180.

[12] Bazerman, Charles. *Shaping Written Knowledge: The Genre and Activity of the Experimental Article in Science.* University of Wisconsin Press: Madison, Wisconsin, 1988. 320.

[13] All assignments are graded by the instructor and passing grades indicate to students they are to proceed to the next assignment. Although student groups

are not required to formally submit their research to their identified audience, some groups have decided to share their findings.

[14] Sheehan, Richard J. and Andrew Flood. "Genre, Rhetorical Interpretation, and the Open Case: Teaching the Analytical Report." *IEEE Transactions on Professional Communication*, Vol. 42, No. 1, 1999, 23.

[15] Spinuzzi, Clay. "Psuedotransactionality, Activity Theory, and Professional Writing Instruction." *Technical Communication Quarterly* 5, no. 3 (1996): 295.

The term "psuedotransactionality" was originally coined by Joseph Petraglia. See: Petraglia, Joseph. "Spinning Like a Kite: A Closer Look at the Pseudotransactional Function of Writing." *Journal of Advanced Composition* 15 (1995): 19-33.

[16] Connors, Robert J. "Landmark Essay: The Rise of Technical Writing Instruction in America." *Three Keys to the Past*. Eds. Teresa C. Kynell and Michael G. Moran. Samford, CT: Ablex Publishing, 1999. 189.

[17] Connors explains that prior to the formation of a technical writing discipline, "there was no glory and no real chance for professional advancement" if an English professor decided to pursue teaching communication to engineers. It was, therefore, assumed that if a professor was teaching technical writing, he or she was forced into the position and perceived as "second rate." Connors, Robert J. "Landmark Essay: The Rise of Technical Writing Instruction in America." *Three Keys to the Past*. Eds. Teresa C. Kynell and Michael G. Moran. Samford, CT: Ablex Publishing, 1999. 178.

[18] Case Western Reserve University English Department "English Graduate Concentration in Writing History and Theory" <http://www.case.edu/artsci/engl/html/whit.html>

[19] Harris, Randy Allen. *Landmark Essays on the Rhetoric of Science*. Mahwah, N.J: Hermagoras Press. 1997. xii.

[20] Bazerman, Charles. *The Languages of Edison's Light*. Cambridge: The MIT Press, 1999. 4.

[21] P. Bizzell & B. Herzberg (Eds.) *The Rhetorical Tradition: Reading from Classical Times to the Present*. Boston: Bedford/St. Martin's, 2001. 81.

[22] Alan Gross bluntly remarks, "the 'brute facts' themselves mean nothing; only statements have meaning, and of the truth of those statements we must be persuaded." Gross, Alan G. *The Rhetoric of Science*. Cambridge: Harvard University Press, 1996. 4.

[23] Bazerman, Charles. *The Languages of Edison's Light*. Cambridge: The MIT Press, 1999. 85.

[24] Connors, Robert J. "Landmark Essay: The Rise of Technical Writing Instruction in America." *Three Keys to the Past*. Eds. Teresa C. Kynell and Michael G. Moran. Samford, CT: Ablex Publishing, 1999. 192.

[25] Florman, Samel C. *The Introspective Engineer*. New York: St. Martin's Press, 1996. 183.

WORKS CITED

The Accreditation Board for Engineering and Technology web site: http://www.abet.org

Anderson, Paul V. *Technical Communication: A Reader Centered Approach*. Boston: Heinle, 1997.

Anderson, Virginia and Barbara Walvoord. *Effective Grading: A Tool for Learning and Assessment*. San Francisco: Jossey-Bass, 1998.

Aristotle. *On Rhetoric: A Theory of Civic Discourse*. Trans. G. A. Kennedy. New York: Oxford: Oxford University Press, 1991.

Bain, Ken. "What Makes Great Teachers Great?" *The Chronicle of Higher Education* 50.31(9 April, 2004).

Bazerman, Charles. *The Languages of Edison's Light*. Cambridge: The MIT Press, 1999.

Bazerman, Charles. *Shaping Written Knowledge: The Genre and Activity of the Experimental Article in Science*. Madison: University of Wisconsin Press, 1988.

Beer, David and David McMurrey. *A Guide to Writing As An Engineer*. 2nd Ed. New York: John Wiley and Sons, 2004.

Connors, Robert J. "Landmark Essay: The Rise of Technical Writing Instruction in America." *Three Keys to the Past: The History of Technical Communication*. Ed. Teresa C. Kynell and Michael G. Moran. Stamford, CT: Ablex, 1999.

Covington, M. V., and Wiedenhaupt, S. "Turning Work into Play: The Nature and Nurturing of Intrinsic Task Engagement." *Effective Teaching in Higher Education: Research and Practices*. Ed. R. Perry and J.C. Smart. New York: Agathon Press, 1997.

Florman, Samuel C. *The Introspective Engineer*. New York: St. Martin's Press, 1996.

Gross, Alan G. *The Rhetoric of Science*. Cambridge: Harvard University Press, 1996.

Halloran, S. Michael. "From Rhetoric to Composition: The Teaching of Writing in America to 1900." *A Short History of Writing Instruction: From Ancient Greece to Twentieth-Century America*. Ed. James J. Murphy Davis: Hermagoras Press, 1990.

Harris, Randy Allen. *Landmark Essays on the Rhetoric of Science*. Mahwah, N.J: Hermagoras Press, 1997.

Kuhn, Thomas S. *The Structure of Scientific Revolutions*. 3rd Ed. Chicago: Univ. of Chicago Press, 1996.

Lannon, John M. *Technical Communication*. 9th Ed. New York: Longman, 2003.

Markel, Mike. *Technical Communication.* 7th Ed. Boston: Bedford/St. Martin's, 2003.
Rhetoric: Concepts, Definitions, Bounderies. Ed. William A. Covino and David A. Jolliffe. Boston: Allyn and Bacon, 1995.
Sheehan, Richard J. and Andrew Flood. "Genre, Rhetorical Interpretation, and the Open Case: Teaching the Analytical Report." *IEEE Transactions on Professional Communication* 42.1, (1999).
Spinuzzi, Clay. "Psuedotransactionality, Activity Theory, and Professional Writing Instruction." *Technical Communication Quarterly* 5.3 (1996)
Stevenson, Susan and Steve Whitmore. *Strategies for Engineering Communication.* New York: John Wiley and Sons, 2002.
Winsor, Dorothy W. *Writing like an Engineer: A Rhetorical Education.* Hillsdale: Laurence Erlbaum Associates, 1995.

FUTURES

13 The Third Way: PTW and the Liberal Arts in the New Knowledge Society

Anthony Di Renzo

"The knowledge we now consider knowledge proves itself in action. What we now mean by knowledge is information effective in action, information focused on results. The results are seen outside the person—in society and economy, or in the advancement of knowledge itself."

— *Peter Drucker,* Post-Capitalist Society *(46)*

THE THIRD WAY AND THE GLOBAL UNIVERSITY

Four centuries after Sir Francis Bacon first proposed his landmark educational reforms in *The Advancement of Learning* (1605), the liberal arts have become *practice-oriented*, more fully conscious of their concrete application in the marketplace and within public and private institutions. For teachers and scholars of professional and technical writing (PTW), this development represents not only new opportunities for program development but a new model for the humanities themselves. As Richard M. Freeland, the president of Northeastern University, observes: "Slowly, but surely, higher education is evolving a new paradigm for undergraduate study that erodes the long-standing divide between liberal and professional education. Many liberal arts colleges now offer courses and majors in professional fields: professional disciplines, meanwhile, have become more serious about the arts and sciences. Moreover, universities are encouraging students to include both liberal arts and professional coursework in their programs of study, while internships and other kinds of off-campus experience have gained widespread acceptance in both liberal and professional disciplines" (141).

Freeland calls this paradigm the Third Way, but its premises are hardly new. PTW programs have advocated these cross-disciplinary ideas since Carolyn R. Miller's 1979 essay "A Humanistic Rationale for Technical Writing." Mainstream educators, however, largely ignored us, until the Association of American Colleges and Universities (AAC&U) published its 2002 national panel report,

Greater Expectations: A New Vision for Learning as a Nation Goes to College. Over the past four decades, the report documents, college attendance has grown so much that seventy-five percent of high school graduates now get some postsecondary education within two years of receiving their diplomas. This remarkable trend, true not only in America but abroad, has resulted from the latest seismic shifts produced by the Scientific and Industrial Revolutions Bacon foresaw in the early seventeenth century. According to Adrian Woodridge, a Washington D.C. correspondent for *The Economist,* four economic and technical developments have caused this international boom in higher education:

- *Democratization:* Or "massification" in the jargon of the educational profession (3). All over the world, more people than ever are attending college.
- *Globalization:* The "death of distance" is transforming education just as radically as it is transforming the economy (3).
- *Competition:* Traditional universities are being forced to compete for students and research grants, and private companies are trying to break into a sector which they regard as "the new health care" (3).
- *The Rise of the Knowledge Economy:* The world is in the grip of a "soft revolution" in which "knowledge is replacing physical resources as the main driver of economic growth" (3).

This last factor, we believe, is most responsible for the emergence of Freeland's Third Way, and for the growing importance of PTW programs at liberal arts colleges. As the AAC&U observes, students are flocking to college "because the world is complex, turbulent, and more reliant on knowledge than ever before" (1). Ironically, however, educational practices invented when higher education served only the few are increasingly disconnected from the needs of these contemporary students. If the humanities are to remain viable, dynamic, and relevant, the panel concludes, liberal arts colleges must redefine their mission:

> Liberal education for the new century should look beyond the campus to the issues of society and the workplace. It should aim to produce global thinkers. Quality liberal education should prepare students for active participation in the private and public sectors, in a diverse democracy, and an even more diverse global community. It will have the strongest impact when studies reach beyond the classroom to the larger community, asking students to apply their developing analytical skills and ethical judgment to concrete problems in the world around them, and to connect theory with insights gained

from practice. This approach to liberal education—already visible on many campuses—erases the artificial distinction between studies deemed liberal (interpreted to mean they are not related to job training) and those called practical (which are assumed to be). A liberal education is a practical education because it develops just those capacities needed by every thinking adult: analytical skills, effective communication, practical intelligence, ethical judgment, and social responsibility (5).

In the past, colleges grudgingly provided such skills by offering the most basic professional and technical writing courses. But as more campuses practice what William Butcher calls "the applied humanities," more educators have become dissatisfied with these skills-based service courses (624). For the liberal arts to prepare students for "responsible action," for business, communications, health, law, and technologies to become a form of "liberal education," professional training itself must become an object of genuine intellectual inquiry and a topic for serious writing (AAC&U 3). That means developing comprehensive and interdisciplinary PTW programs, housed in and treated as a branch of the humanities, but that serve the professions.

This essay collection attempts to describe the mission, curriculum, and administration of such programs at a dozen liberal arts colleges. Like the science ministers in Bacon's *New Atlantis* (1625), the editors have gathered "books, abstracts, and patterns of experiments" from across the country (MW 486), spotlighting the practical side of Bacon's visionary philosophy. Writing at a time when exploration had opened new trade routes and invention and entrepreneurship had created new technologies, Bacon considered knowledge precious capital in a global market. The parallels to our own time are obvious, but before we discuss the larger implications of these grassroots programmatic developments, let us outline the contents of this book.

PTW PROGRAM DEVELOPMENT AT LIBERAL ARTS COLLEGES

We were surprised to find when surveying proposals and drafts of papers for this collection how many of the authors were new Assistant Professors, inventing and designing programs that were often new to the college or department, and new as well to the director. Of course, many programs have long pedigrees and seasoned faculty, but Professional and Technical Writing programs are proliferating, and it is not possible in a new program to pick up where others left off. Thus the director and colleagues are compelled to make what are often, for that site, decisions without precedent, such as how the program is to be advertised and consequently what students one attracts, the kinds of assessment materials to be solicited and preserved, the sort of relations the PTW program is to have to other departments and faculty, and much more. All this is rather challenging and rewarding, as anyone reading this is likely to agree, and the rewards and challenges come not so much from the enrollment numbers or class assessment forms, but from a deep and fulfilling sense of having contributed to something worthwhile, an institutional structure that elicits curiosity and rigor in the students—and helps them move on with their lives.

THE DIVISION OF KNOWLEDGE AND THE APPLIED HUMANITIES

These developments are crucial to our field, but their implications go far beyond it. They involve such larger issues as post-modernity, globalization, and mass democracy, and their impact on the liberal arts. If traditional humanism is to survive, it must come to terms with thinking and communicating within a high-tech, commercial society. Over a decade ago, Gerald Graff in *Beyond the Culture Wars* advocated "teaching the conflicts" as a way to "revitalize" higher education, but when it comes to the new corporate university, most humanists still indulge in "apocalyptic posturing" (5). We denounce the military-industrial-academic complex while conducting online research and applying for institutional grants. By denying our paid function within a post-capitalist economy, we alienate our students, who attend college primarily to become knowledge workers, and ignore a watershed cultural development: the changing nature and role of disciplinary and professional knowledge in our time.

As Nobel economist Friedrich Hayek noted seventy years ago, the division of labor, which made possible the Industrial Revolution, has become a more subtle "*division of knowledge,*" which now characterizes and sustains modern business, political, and academic institutions (50). Borrowing a term from Adam Smith, Hayek called our knowledge-based civilization the Great Society: social arrangement based on widespread and decentralized economic interdependence, abstract legal codes, and impersonal information rather than local and concentrated family ties, concrete tribal customs, and personal dialogue. The division of knowledge, therefore, carries profound political, ethical, and rhetorical significance. As Hayek declares in "The Use of Knowledge in Society:" "We make constant use of formulas, symbols, and rules whose meaning we do not understand and through the use of which we avail ourselves of the assistance of knowledge which individually we do not possess" (88). Since no one possesses total knowledge, Hayek concludes, different disciplines and professions of knowledge must learn to understand and dialogue with each other.

In *Post-Capitalist Society,* Peter Drucker shows how Hayek's theory affects both business and the academy. Underlying all three phases in the global shift to a knowledge economy—the Industrial Revolution, the Productivity Revolution, the Managerial Revolution— has been a fundamental shift in the meaning of knowledge itself: "We have moved from knowledge in the singular to knowledges in the plural," Drucker explains (45). Traditional knowledge was holistic and general; contemporary knowledge, in contrast, is partitioned and highly specialized, focused on practice and concerned with results. "This is as great a change in intellectual history as ever recorded," Drucker declares (46). While the traditional university demoted specialized knowledges to the level of "crafts," the modern university elevates them to "disciplines" and "professions" (46). Such privileging is fitting, Drucker argues, for without this necessary specialization of disciplines and professionals, mass society and the global economy would collapse, and billions would perish.

> The shift from knowledge to knowledges has given knowledge the power to create a new society. But this society has to be structured on the basis of knowledge as something specialized, and of knowledge people as specialists. This is what gives them their power. But it also raises basic questions—of values, of vision, of beliefs, of all the things that hold society together and give meaning to our lives. . . . [I]t also raises a big—and new—question: what constitutes the educated person in the society of knowledges? (46-47)

According to PTW scholar Bernadette Longo, this question originates with Sir Francis Bacon, who "coined" the concept of a practice-oriented academy and "minted" the discipline of professional and technical rhetoric (21). Surveying the progress made in the early seventeenth century, he compared the New Learning to a galleon returning through the Straits of Gibraltar, loaded with the bounty of invention and enterprise from foreign ports. Unfortunately, Bacon complained, such cargo was warehoused on a rotting pier. Rather than circulate knowledge, the academy of Bacon's day hoarded it in dry dock. Bastions of power and privilege, Oxford and Cambridge had built moats to contain new currents of thought and had become fortified worlds unto themselves. Rather than face and ponder the implications of new markets and technologies, England's best universities idolized the past, disdained the present, and feared the future. In addition, traditional scholars and rhetoricians were obsessed with words, not things, while skeptical philosophers and jaded historians promulgated "the doctrine of Acatalepsy," the radical belief that all human knowledge ultimately is impossible (NO 75). In the name of Humanism, Bacon accused, both practices betrayed humanity by traducing reason. "The first subdues the understanding," he observed, "the second unnerves it" (76).

For PTW programs, this stalemate depressingly resembles the vicious culture wars within the current humanities between traditionalists and theorists, which have done so much to discredit the liberal arts in the eyes of students and the public. Echoing C.P. Snow, John Brockman, author of *The Third Culture: Beyond the Scientific Revolution,* harshly criticizes this state of affairs:

> American intellectuals are, in a sense, increasingly reactionary, and quite often proudly (and perversely) ignorant of many of the truly significant intellectual accomplishments of our time. Their culture, which dismisses science and industry, is often non-empirical. It uses its own jargon and washes its own laundry. It is chiefly characterized by comment on comments, the swelling spiral of commentary eventually reaching the point where the real world gets lost.

In contrast, the Third Culture, Brockman's term for the scientific and technical culture of the professional knowledge society, can tolerate disagreements about which ideas to take seriously precisely because its diverse specialization recognizes no canon or accredited list of acceptable ideas. More to the point, it reaches beyond the academy. Since its members communicate effectively not only to each other but to legislators, business leaders, the media, and the public, its ideas have greater currency. Unlike past intellectual pursuits, therefore, the Third Culture's achievements "are not the marginal disputes of a quarrelsome mandarin class: they will affect the lives of everybody on the planet."

Bacon's warning in *The Great Instauration* (1620), therefore, remains relevant. Puffed by the winds of pride and rocked by the storms of controversy, the traditional liberal arts veer close to shipwreck. To chart a better course, disciplines must abandon dead reckoning and rely on compass and quadrant, if only to draw more accurate maps and to train more expert navigators. Likewise, the humanities should emphasize the practical application of knowledge, should confront science and economics and integrate technology within their curricula, and should dedicate themselves, at least partially, to professional training and institutional administration.

PTW, *RHETORS,* AND THE FUTURE OF THE LIBERAL ARTS

For practical and theoretical reasons, however, college administrators must honor and support composition and PTW programs if the applied humanities are to succeed. Since the dawn of the Scientific and Industrial Revolutions, writing has remained an essential skill in the marketplace, resulting in the mass literacy of the nineteenth and twentieth centuries, but the dialogical nature of writing itself also teaches how knowledge emerges and circulates within Adam Smith and Friedrich Hayek's Great Society. Observing the great knowledge shift in his own day, Francis Bacon remarked: "Up to now, thinking has played a greater part than writing in the business of invention, and experience has yet to become literate. But no adequate inquiry can be made without writing, and only when that comes into use and experience learns to read and write can we hope for improvement" (109).

Accordingly, Bacon dedicated large sections of his revised edition of *The Advancement of Learning* to composition and professional and technical writing. These subjects remain crucial to the academy's identity and survival in the new Knowledge Society. Composition teaches students how to mediate between the competing discourse communities of different academic disciplines, while professional and technical writing teaches students to mediate between the competing discourse communities of different professions. The first demonstrates how disciplines profess knowledge, the second how knowledge disciplines the professions. This dialectic, Peter Drucker insists, is crucial in a world defined and determined by the division of knowledge:

> We neither need nor will get "polymaths" who are at home in many knowledges; in fact, we will become even more specialized. But what we do need—and what will define the educated person in the knowledge society—is the ability to *understand* the various knowledges. What is each one about? What is it trying to do? What are its central concerns and theories? What major new insights has it produced? What are its important areas of ignorance, its problems, its challenges? (217)

Such interdisciplinary cross-pollination allows new ideas to flower, both in the academy and the marketplace. Bacon addresses this issue in his famous fable in the *Novum Organon*. Playing entomologist, Bacon divides knowledge workers into three kinds of insects. Ants blindly collect and use facts; spiders spin webs of sophistry from their butts; but bees gather material from the flowers of the garden and field, then transforms and digests it by an internal process. "And the true business of philosophy is much the same," Bacon concludes, "for it does not rely chiefly on powers of the mind, nor does it store the material supplied by natural history and practical experiments untouched by memory, but lays it up in the understanding changed and refined (105). Distilled and circulated, the nectar of knowledge creates the honeycomb of Adam Smith's

Great Society. In the resulting buzz of the marketplace, citizens "must practice oratory" their entire lives, Smith maintained, trading words and expertise as well as goods and services to promote their individual good, and the general welfare of the hive (Fleischacker 92).

Given this social reality, shouldn't the academy teach students to become effective and ethical *rhetors* in an emerging knowledge economy, particularly when globalization and technology have brought our planet to a historic turning point? Until recently, the liberal arts have ducked this question as matter of principle, supposedly because techno-capitalism is inherently dehumanizing and because humanists should dissuade students from seeking professional training. Matthew Arnold, the apostle of the Great Tradition, took a similar position in his famous debate with Thomas Huxley, who called for a "practical" liberal education at a time when science and industry had dramatically increased access to public schooling in Victorian England (244). Nevertheless, Arnold conceded a major point:

> [The traditional liberal arts] show the influence of a primitive and obsolete order of things, when the warrior caste and the priestly caste were alone in honor, and the humble work of the world was done by slaves. We have now changed all that; the modern majesty consists of work, as Emerson declares; and in work, we may add, principally of such plain and dusty kind as the work of cultivators of the ground, handicraftsmen, men of trade and business, men of the working professions. Above all is this true in a great industrious community such as that of the United States. (75)

Plato's Academy is not fit for our world, Arnold admitted, because Plato, who scorned handicraft and the professions, never could have foreseen a capitalist society. "Such a community must and will shape its education to suit its own needs," Arnold said. "If the usual education handed down to it from the past does not suit it, it will certainly before long drop this and try another" (76).

Like Dr. Arnold, contemporary advocates of the liberal arts should be gracious and perceptive enough to see consumer demand for professional training in higher education as an expression of mass democracy, and an equally valid form of humanism. As Fr. Walter Ong pointed out in his 1978 MLA President's Address, markets and technologies increasingly attract college students, "not because they are inhuman, but because they are eminently human, the creations of human beings" (1916). Like language itself, they are media of exchange and as such deserve scholarly respect and attention. As their very title implies, professors are professionals, too, Ong reminded his fellow scholars, with "fiducial" responsibilities to their institutions and clientele (1911).

Richard Freeland agrees. "Claims for the moral superiority of liberal education reflect a bias against—even a disdain for—the workaday earning experiences of most adults," he states, "as if academic learning had a monopoly on value and meaning and other forms of work were solely about material gain. This perspective is an unfortunate relic from the tradition of classical—and class-based—education in Britain, from which the contemporary liberal arts are descended" (147). Following the counter-tradition of Bacon, Smith, and Huxley, Freeland argues that professionalism and humanism, with the proper education and under the right conditions, can and should be synonymous:

> Instead of deriding students' interest in their careers, we should help them see how the work they do can promote personal growth, intellectual adventure, social purpose, and moral development. We should show them how the values of intellectual honesty, personal integrity, and tolerance can strengthen the institutions in which they will work. And we should help them build bridges between the intellectual concerns they encounter in philosophy, literature, and history courses and the decisions they will have to make as business leaders, lawyers, and government officials. Properly conceived, practice-oriented education can provide at least as powerful a moral education as any purely academic study of ethics. (147)

Committed to practice-oriented education, professional and technical writing programs are vanguards of the Third Way, helping colleges as well as students grapple with current political and economic realities. For disaffected hardliners on the left and right, this development represents the academy's surrender to corporate values; but for the more open-minded and engaged, professional and technical writing programs provide a way to take advantage of the genuine benefits associated with the emerging global university. "There are plenty of justifications for the revolution that is sweeping through higher education," Adrian Woodridge suggests:

> It is giving students more control over where they get educated. It is giving millions of youngsters a chance to study abroad. It is throwing up colleges that can teach managerial and technical skills. It is reconnecting academics with the wider knowledge economy. But the most important justification of all is that it is freeing resources for intellectual activity. It is filling libraries

with books, stocking laboratories with equipment, and giving more researchers than ever before a chance to produce order out of chaos. (22)

Whether these benefits ultimately outweigh the drawbacks depends on vision, wisdom, and action. As globalization drags the liberal arts, kicking and screaming, from the cloister to the market, PTW programs can prepare the academy to face its greatest contemporary challenge. With compelling urgency, Peter Drucker describes the stakes:

> The knowledge society *must* have at its core the concept of the educated person. It will have to be a universal concept, precisely because the knowledge society is a society of knowledges and because it is global—in its money, its economics, its careers, its technology, its central issues, and above all, in its information. Post-capitalist society requires a unifying force. It requires a leadership group, which can focus local, particular, separate traditions into a common and shared commitment to values, a common concept of excellence, and on mutual respect. (212)

The interminable debates between theorists and humanists, therefore, are a dead end. Global society needs the very thing radical skeptics reject: a universally educated person. At the same time, the great Western tradition, which humanists defend, is inadequate for a postcolonial world. Humanists can offer only a bridge to the past, when students need to bring their knowledge to bear on the present with the hope of shaping the future. Without that practical application, as Peter Drucker observes, humanist values are "only fool's gold unless they have relevance to the world" (213). Professional and technical writers have known this truth since Agricola. Real gold must be mined, smelted, and coined, and PTW programs can provide colleges with the rhetorical tools—the practical and intellectual tools and techniques—to forge a new humanism suitable for the perils and promises of a new century.

WORKS CITED

Association of American Colleges and Universities. "Executive Summary." *Greater Expectations: A New Vision for Learning as a Nation Goes to College*. National Panel Report, 2002: 1-5.

Arnold, Matthew. "Literature and Science." *The Harper and Row Reader: Liberal Education Through Reading and Writing*. 3rd Ed. Ed. Marshall Gregory and Wayne Booth. New York: Harper, 1992. 73-89.

Bacon, Sir Francis. *The Major Works*. Revised. Ed. and Intro. Brian Vickers. Oxford: Oxford University Press, 2002.

Bacon, Sir Francis. *Novum Organum*. Ed. and Trans. Peter Urbach and John Gibson. Chicago: Open Court Press, 1994.

Brockman, John. "The Third Culture." Edge Foundation., Inc. 2001. 20 March, 2006 <http://www.edge.org/3rd_culture/index.html>

Butcher, William F. "Applied Humanities: They Will Pay You for the Other Five Percent." *Vital Speeches of the Day*. (1 August, 1990): 623-625.

Drucker, Peter F. *Post-Capitalist Society*. New York: Harper Collins, 1994.

Fleischacker, Samuel. *On Adam Smith's* The Wealth of the Nations. Princeton: Princeton University Press, 2004.

Freeland, Richard M. "The Third Way." *The Atlantic Monthly*. (October 2004): 141-147.

Graff, Gerald. *Beyond the Culture Wars: How Teaching the Conflicts Can Revitalize American Education*. New York. W. W. Norton, 1992.

Hayek, Friedrich A. *Individualism and Economic Order*. Chicago: University of Chicago Press, 1980.

Huxley, Thomas Henry. "Science and Culture." *Classical Rhetoric for the Modern Student*. 4th ed. Ed. Edward Corbett and Robert Connors. Oxford: Oxford University Press, 1999. 237-45.

Longo, Bernadette. *Spurious Coin: A History of Science, Management, and Technical Writing*. Albany: SUNY Press, 2000.

Ong, Walter. "MLA Presidential Address, 1978: The Human Nature of Professionalism." *PMLA: Special Millennium Edition*. 115.7 (December 2000): 1906-1917.

Smith, Adam. "Of the Expense of the Institutions for the Education of Youth." *The Essential Adam Smith*. Ed. Robert Heilbroner. New York: Norton, 1986. 303-307.

Woodridge, Adrian. "The Brains Business: A Survey of Higher Education." *The Economist* (10 September, 2005): 3-22.

14 The Write Brain: Professional Writing in the Post-Knowledge Economy

Alex Reid

The inclusion of computer technology in writing degrees is hardly new. Indeed it is the hallmark of technical writing degrees. While the history of technical writing follows the rise of our industrial economy, technical writing is a prototypical career of the *post-industrial*, knowledge economy. Technical writers—like the engineers, computer programmers, lawyers, accountants, and other experts whose knowledge they translate—have played an important role in the professional world of the last thirty years. However, today, the jobs of the knowledge economy, like industrial jobs before them, are moving overseas. While these jobs will not disappear overnight, in designing a professional writing curriculum, it is important to anticipate the changing requirements of our field as we move toward a "post-knowledge" economy. As I describe in this chapter, this future is one that will require not only solid technical skills but also strong creative and rhetorical abilities to empathize with, and design powerful experiences for, a variety of audiences/users. In this context, we have built our professional writing curriculum partly upon the traditions of technical writing, while also drawing from creative writing and journalism and more generally from the discipline of rhetoric and composition. We have also looked beyond our discipline for the teaching of media production, multimedia design, and information management.

In doing so however, we find ourselves pressured from two ends. We foresee a marketplace seeking a more sophisticated, creative, and technologically proficient writer, but we encounter incoming students with an increasing need for instruction in what we have traditionally viewed as fundamental skills in writing. In this situation, we cannot simply add new curriculum onto our existing professional writing curriculum. To do so would establish an expanding curriculum with escalating credit hours (something undesirable for our students and increasingly unmanageable for a small faculty). Instead, we find ourselves with the task of building a curriculum that blends the emerging expectations of professional writing with more traditional models of technical communication and advanced writing genres. Rather than thinking of creative writing, technical writing, other genres, or general composition as discrete subjects (or even majors), we have found it necessary to conceive professional writing at the intersection of these and other writing traditions with emerging rhetorical concerns in design, information, and multimedia production. In doing so, we find ourselves confronting some fundamental notions about writing and writing instruction, a confrontation I suppose should not be unexpected given the dramatic changes in media, information, and communication we are experiencing.

In discussing these challenges, I have divided this chapter into three main parts. In the first part, I address in greater detail this emerging post-knowledge economy, which Daniel Pink terms the "Conceptual Age." Pink employs the familiar shorthand of left- and right-brain functions to describe how we are moving from a left-brain oriented knowledge economy to a new economy that will place greater emphasis on the creativity and empathy associated with the right hemisphere of the brain. My interest here is in identifying how such a shift might inform the development of professional writing. In the second section, I turn specifically to the role of Web 2.0 technologies in our curriculum. Web 2.0 technologies, such as blogs, wikis, and social bookmarking sites, are an important part of the economy Pink and others describe. Here I discuss how industry concepts of Web 2.0 practices might apply to building a curriculum. Of course, one of the primary challenges to such an adoption is keeping faculty current with emerging technologies. I address this subject in the chapter's third section where I consider the viability of bottom-up approaches for adopting technologies in an academic context. While I am particularly focused here on current technologies, I also want to emphasize that our disciplinary goal should not solely be how to integrate these specific applications but also how to create curricular structures and practices that will allow us to deal on an ongoing basis with emerging technologies.

WRITING, KNOWLEDGE WORKERS, AND THE RIGHT-BRAIN

The metaphor of "left-brain" and "right-brain" functions, attitudes, and proclivities has become fairly common in our culture. While cognitive science does identify the different hemispheres of the brain as having different functions, in general our daily activities involve both sides to one degree or another. That said, the *metaphor* of left and right sheds light on how we view and value different cognitive functions within our culture. As Daniel Pink notes in *A Whole New Mind*, the post-industrial era has emphasized "left-brain" cognitive skills as the foundation of our knowledge economy. This left-brain orientation is characteristically "sequential, literal, functional, textual, and analytic" (26). It typifies the type of work traditionally done by engineers and computer programmers but also by lawyers, accountants, radiologists, MBA's and many other professionals. Certainly technical writers would fall into this category. Pink's characterization of left-brain thinking would make a reasonable description of the values of technical communication. In comparison to other genres of writing, the emphasis of technical writing has always been on clear, structured, logical, and rational communication. In turn, technical writing courses and programs have emphasized the development of writing skills along those lines. The result, in general, is that graduates of technical writing programs develop a complex set of rational and analytic cognitive abilities, much like their knowledge worker colleagues in law, engineering, computer science, and so on.

While skilled knowledge workers remain very much in demand, Pink and many others have noted an increasing trend that will likely alter that demand in significant ways. The phenomenon of "outshoring," the exporting of knowledge worker jobs to Asia, has been extensively reported, if not over-hyped, in recent years. However, in the long term (though certainly within our students' professional lives), jobs in the knowledge economy will likely meet a fate similar to that which jobs in the industrial economy met a few decades ago. In addition to the exportation of jobs to Asia and elsewhere, the increasing sophistication and power of computers allows them to undertake many of the fundamental functions performed by knowledge workers. As Pink notes, "the Web is cracking the information monopoly that has long been the source of many lawyers' high income and professional mystique. Attorneys charge an average of $180 per hour. But many Web sites—for instance, Lawvantage.com and MyCounsel.com—now offer basic legal forms and other documents for as little as $14.95" (46). Obviously a web site isn't going to replace all lawyers, but clearly many, many lawyers, especially junior lawyers, earn a living performing relatively simple legal tasks or conducting research, which either can be done by a computer or more cheaply by a knowledge worker living in Asia. While I don't believe computers will be writing their own documentation any time soon, our ability to discover and share information over networks is altering the way technical communication is done.

Ultimately Pink's argument is not that left-brain thinking will not be valued in the future. Instead, as his book's title suggests, he simply sees a rising appreciation for right-brain thinking to the point where future careers in America will require a *whole* new mind, both left and right. In distinction from left-brain thinking, Pink describes right-brain thinking as "simultaneous, metaphorical, aesthetic, contextual, and synthetic" (26). In other words, "right-brain" activities are those that allow us to "see the big picture," to incorporate intuitive or empathic understanding, to make connections between ideas that are logically unrelated, and to process the complexities of embodied experiences without relying upon abstractions. In terms of writing instruction, if the left-brain reflects the emphases of technical writing (and to a lesser degree, composition), then the right-brain connects to some of the traditional values of creative writing. One might be tempted to go so far as to suggest that rhetoric is left-brain and poetics is right-brain. However, while there may be some validity to that suggestion, at least in terms of how these issues have played out within disciplinary politics, I contend, following Pink at least this far, that successful writing requires a whole mind, particularly as we prepare students for the demands of careers in the post-knowledge economy.

Jon Udell, writing for the O'Reilly Network, picks up on this shift in the values relating to professional preparation and goes so far as to suggest that the future of first-year composition will be characterized by the production of multimedia documents, which he terms screencasts.[1] Udell sees screencasts as being rhetorically different from traditional genres of professional writing related to software development, which might be divided into technical or support documents and marketing or sales materials. The purpose of the screencast will be to connect end-users with the designers of new technologies and applications. Udell writes, "the rate-limiting factor for software adoption is increasingly not purchase, or installation, or training, but simply 'getting it'... We haven't always seen the role of the writer and the role of the developer as deeply connected but, as the context for understanding software shifts from computers and networks to people and groups, I think we'll find that they are" (2005). In short, as information technologies become increasingly about social uses (e.g. wikipedia, del.icio.us, flickr), there is an increasing need for writers who can communicate the social dynamics of a technology; that is, someone who will be able to work with developers in helping to articulate and communicate their vision. As Udell continues, "*The New York Times* recently asked: 'Is cinema studies the new MBA?' I'll go further and suggest that these methods ought to be part of the new freshman comp. Writing and editing will remain the foundation skills they always were, but we'll increasingly combine them with speech and video. The tools and techniques are new to many of us. But the underlying principles—consistency of tone, clarity of structure, economy of expression, iterative refinement—will be familiar to programmers and writers alike."

Udell's vision may still sound very much, in principle, like the traditional values of technical composition, plus the addition of new media, in that he makes reference to values like consistency, clarity, economy, and refinement. However, there is a deeper transformation taking place in the coming together of media and the identification of a new purpose and new audience, specifically in Udell's suggestion that screencasts need to help their audience "get it," to see the social value of a new application. "Getting it" is not particularly a matter of rationally communicating the various features of an application (as technical documentation would) or even selling those features or some feeling a company hopes to associate with an application (as marketing or advertising media would). Instead, Udell describes an emerging genre that seeks to demonstrate to potential users the ways in which a new application might fit into lives and allow them to make better use of the increasing amount of media available to them. For example, it is not enough for the developers of blogging applications to provide technical documentation or to produce advertising for their service; they need to communicate to potential users how a blog will allow them to participate in a community of readers and writers. This participation gets more specific as one thinks about particular types of bloggers: educators using blogs in their classes, professional writers who want to make money from their blogs, companies using blogs for internal communication or to communicate with clients or to market products, and other individuals who simply wish to keep a diary or share a personal interest or viewpoint. Of course, the audiences become even more specific than that (e.g., addressing the use of blogs in first-year composition courses). A screencast for blogging in composition would include video, audio, and text that would demonstrate how you might easily set up a blog to share information with students, to have students comment on readings, or to distribute and comment on drafts of more formal writing assignments; it might also discuss how giving students the experience of producing their own blog creates an opportunity for investigating how discursive practices and a sense of audience develop in a new medium. Whatever the particular content of the screencast, the basic point is that it requires a new rhetorical, compositional approach in which writers and developers strive to help their potential users see how a new application fits into a larger picture of their information habits.

This shift away from instrumental reason is echoed elsewhere in the rethinking of professionalizing education. Richard Gabriel, a Distinguished Engineer at Sun Microsystems, has argued that software engineering programs should pattern themselves after MFA programs in creative writing. In particular, Gabriel references the system of mentoring, the community of writers, and the curriculum of ongoing practice, reflection, and revision in the context of workshops, conferences, and other coursework. He recognizes that "in software as in writing, there are people whose work is 'doing the thing'—writing and designing programs—and such people do this work every day. They hope to be good at it and to be able to improve over time. They have pride in what they do and are satisfied or not with each project they do. To them what they do feels more than craft, includes engineering and science, but still feels like more." Gabriel is articulating the need for software engineering programs to develop a reflective and broader vision of the process of composition, one that does not focus solely on the grammar of coding or other practical issues but, as Udell is suggesting, aids software designers to develop an ability to "get it" and communicate their understanding to others. In many ways this is much like an MFA program that assists writers in developing a critical understanding of their own writing. That is, it is one thing to have some native sense of when one's writing is or isn't working; it is another matter to develop the critical ability to explain why a piece of writing does or doesn't "work."

Pink articulates this shift in education in terms of a demand for professionals with an understanding and appreciation of design. As he observes, "Getting admitted to Harvard Business School is a cinch. At least that's what several hundred people must think each year after they apply to the graduate program of the UCLA Department of Art—and don't get in…A master of fine arts, an MFA, is now one of the hottest credentials in a world where even General Motors is in the art business…the MFA is the new MBA" (54). As Pink explains, the growing interest in design comes in part from the incredible abundance and range of choice consumers encounter. Increasingly the primary difference between products is their design. While to a certain extent the process of design relies upon left-brain thinking and scientific knowledge, it is also clearly a right-brain activity dependent upon an appreciation for aesthetics and an intuitive understanding of user experience. In the world Pink, Gabriel, and Udell describe, it will not be sufficient for a professional writer to produce clear and rational prose. Instead, the demand will increasingly be for professional writers who can *also* contribute to user experiences through aesthetics, empathy, narration, and so on. In designing a product and the documentation that might accompany it, a writer must not only clearly communicate the product's functionality but also assist the user in imagining meaningful purposes and creating positive user experiences. This would be the case whether the writer is producing text and media that support or market a non-textual product or if the product is a piece of media itself.

There are many different ways to approach the issue of design. In emphasizing the MFA, Pink identifies the Art orientation, which would include graphic design and other commercial art. There is also product design, designing graphical user interfaces (GUIs), and architecture to name some obvious other examples. The difference is that design in an artistic sense can often be quite distanced from any functionality, particularly in comparison to the relation between design and function in the other examples I provided. Certainly, a graphic element may be called upon to communicate some information (for example a sign) but often the communicational goals of design elements are more vague (e.g., they might convey a mood). From the perspective of professional writing, design is both an aesthetic and rhetorical concern. Indeed, in the past, rhetoric has been characterized, often with pejorative intent, as "mere ornamentation." With the rise of right-brain thinking, that notion of rhetoric as ornament, as a design strategy, as the practice of shaping user/audience experience, comes into its own. Again, I would reiterate that we would not want to view rhetoric solely in these terms, but the perception of rhetoric as design clearly offers a way to connect rhetoric with the emerging economy. It also offers a way to connect rhetoric with more aesthetic and poetic writing practices *and* informs the intersection of conventional, print, writing instruction with instruction in new media composition.

Of course the right-brain isn't simply about design. Pink lists five other right-brain "senses:" story, symphony, empathy, play, and meaning. Without going into depth about each one, an important underlying ability here is to take information and experience and make connections that are not simply "logical" but that resonate in more immediate and intuitive ways with others. Like design, these all connect directly with rhetorical concerns. A technical document may provide a reader with all the facts, but a story may convey the same purpose in a more meaningful and memorable way. One can arrange information logically into various categories, but it may be more powerful to bring these elements together, to compose them symphonically. Likewise to connect empathically, to provide an openness that invites a playful engagement with possibilities, and to recognize the potential meaningfulness of a concept: these are all significant elements of a rhetorical sense of audience. In short, the right-brain cannot simply be about design without also considering what that design allows us to do, without imagining how a powerful user experience goes beyond the immediate aesthetics and into a more lasting meaning. As such, incorporating these concerns into a Professional Writing program has to be about more than issues of usability or designing media that attracts attention or looks "cool." There has to be a connection from design to communication practices that not only manages to convey information logically and rationally but also connects with audiences in deeper and more meaningful ways. Ideally, one moves from viewing writing as the production of discrete, limited bodies of information to recognizing composition as linking into, shaping, and participating with larger flows of media and experience. The abilities to see this space, to operate within it, and to bring others to it ultimately characterize the right-brain.

WEB 2.0 IN THE PROFESSIONAL WRITING CURRICULUM

Undoubtedly, these flows I am describing have become far more palpable with the emergence of the web and networked, multimedia communication. Certainly computers have heightened our sense of design (e.g., thirty years ago, few people had much sense of what font was). Much of the demand that Pink foresees stems from the need to create meaningful experiences of media. For some, this connection between computers and the right brain might be jarring. Computers have long been associated with left-brain careers. Ideologically and culturally we tend to associate classic right-brain types with a degree of Luddism: the poet, the painter, and so on. People who consider themselves to be weak at math or not particularly interested in science or other traditional left-brain areas also might express trepidation, antipathy, or at least disinterest in computers. In large part this has been because, at least historically, engineers and programmers have designed computers for other engineers and programmers, with little thought for other types of users. However, the rise of the Internet over the last ten years has produced two inter-related types of software that do not fit into the tradition image of the solitary computer and computer user: Social Software and, more recently, Web 2.0 applications. The integration of these technologies into Professional Writing serves several key ends. Most obviously, the students develop fluency with the contemporary operation of the Internet. More importantly, students find themselves confronted with a richer rhetorical environment. They must write to multiple audiences, organize continual flows of data, and compose with layers of media. These challenges ask students to combine left- and right-brain capabilities. They need to learn and use computer applications, organize information, and often communicate complex concepts; they also need to connect empathically with their audiences, integrate text with other media, and operate with an understanding of the larger picture.

While all these ends can be achieved without these technologies, Social Software and Web 2.0 applications help users share information easily and increase the value of the information they share by providing easy ways to organize and search that information.[2] Social Software references technologies that enable "many-to-many" communications, which therefore might include MUDS, MOOS, Internet Relay Chat (IRC), and Instant Messaging, as well as newer technologies, such as social bookmarking (e.g. del.icio.us), blogs, and wikis. These latter technologies also fall into the category of Web 2.0, which also includes applications that are less directly "social," such as Google Maps. The definition of Web 2.0 is half technical (referencing the use of newer approaches to the web such as APIs, AJAX, and RSS)[3] and half marketing (as companies scramble to associate themselves with the buzz). For Professional Writing, the most interesting applications are clearly those that deal directly with the production of text and other media. However, others cannot simply be ignored. For example, Google Maps points to the developing phenomena of the geo-tagged web, an Internet mapped onto physical space through the use of GIS coordinates. This geospatial web has obvious uses for cars and pedestrians equipped with GIS devices, and certainly such a web will require text, a new kind of topography that will help users in understanding the value of such information and imagining how they will incorporate it into their lives. As Udell stresses, increasingly the success of technology relies not simply on rational functionality but the ability of consumers to "get it," to see the value of a product in their lives. And as Pink continues, getting it is part of the larger task of designing user experiences.

Much of the discussion of Web 2.0 deals with commercial concerns, essentially addressing how these emerging applications can be monetized or how they alter business practices or corporate culture. However, with some thought, many of these discussions apply to curriculum development. One of the key points regarding Web 2.0 has been the emphasis on trusting users, both end-users/customers and employees working to adopt new technologies into the workplace. An important part of this trust has been the value that users contribute to the experience of the application over time. This can be seen in popular Web 2.0 sites like del.icio.us, flickr.com, and Wikipedia. The more material that users contribute, the more ways they find to make use of application features, and the more data they provide for organizing the media on the site, the more valuable and useful the application becomes. This development of valuable information helps to create a new market for products on what Chris Anderson has termed "the long tail."[4] While the long tail suggests the possibility of building a new marketplace, it relies upon trusted users sharing information so that users can not only find the products they desire but other products in which they might be interested (Amazon attempts this when it shows users other products viewed or purchased by others). These three qualities—trusting users, developing the value of user contributions, and the long tail—provide some important insight into the role Web 2.0 can play in developing Professional Writing curriculum for the emerging economy.

As one can imagine, it can be difficult to trust users. A manager might have an impulse to control the way his or her employees make use of a new technology: it should be used only for serious business…no personal emails, for example. Similarly, a website might control how users make use of its features or information: that book belongs in science fiction, not mysteries. Clearly, faculty regularly struggle with controlling how students use technology in the classroom (witness the long list of rules that might accompany a syllabus for a class in a computer lab). Other faculty might decry, "no internet sources" for research papers. Public schools limit access to blogs, Facebook, MySpace, and so on. Even faculty are warned against blogging by publications like the *Chronicle of Higher Education*.[5] Trusting faculty and students to use these technologies may be the most difficult step that colleges need to make. Obviously there have been and will be missteps along the way as users organically develop rhetorical practices appropriate for these spaces. This necessity for trust falls not only on institutions. Faculty must trust their colleagues and students, and students must trust their peers and instructors. The real curricular value of these technologies will only emerge as we use them to share information across courses rather than restricting it within the boundaries of a single semester.

For example, I am teaching a class on technical writing and my colleague is teaching one on the history of rhetoric. Both of us want to address the subject of ethos. Clearly we have different contexts and purposes for doing so. However, if we share bookmarks, then we double our resources. We might end up discussing the same web article for different purposes and from different perspectives. It's quite likely that we may share students; these students will have an opportunity to experience some of the key issues about audience and purpose that we regularly discuss in all our courses. That is, they will witness how the same article discussed with different groups of people, in different course contexts, and with different professors leads to significantly different outcomes. Perhaps we encourage and require our students to contribute as well. Over time we develop a healthy and dynamic list of web resources on ethos. Taking this one step further, perhaps we have a wiki for our program. All the courses contribute to it to some degree, and our students rely upon it as a reference as they move through the major. Naturally, my colleagues and I disagree with one another from time to time, and we try to work out these disagreements and represent them on the wiki. Our students also disagree with us and with one another. The wiki becomes a map of our dissensus. However, in order to do it, we have to trust one another and our students. Maybe our students start to use the wiki as a place to publish their poetry or to talk about other aspects of their college lives. The faculty could object because it was our intention to have the wiki be academic. Or, we could trust our student-users and recognize the importance of having our students see the wiki as a community space that they regularly use.

I have already slid from my point about trusting users into recognizing the value of the information they contribute. The real value to a Web 2.0-based curriculum only appears over time. After a few semesters, the shared links, wiki entries, and blog posts begin to accumulate. Material is revisited and revised as courses are offered for a second or third time. The advantage of using a folksonomic[6] approach, where users tag media with context-relevant descriptors, is that one can chart shifts in interests and discourse over time in a program. For example, students tag websites that provide information about careers that interest them. This way one might track a growing interest in the publishing industry or attending graduate school or technical writing or wherever student interests might lead. This type of information can be invaluable in trying to understand our program, and it is produced organically and dynamically by the students rather than through some staid questionnaire. As valuable as this might be for faculty, it is potentially more valuable for students. As a student, one no longer needs to rely solely on individual memory and saved notebooks. The material record of a course is available and searchable. It collects not only one's own contributions but also those of other students and faculty. It also includes the contributions of students and faculty from other semesters. Needless to say, the material produced by students and faculty in one college then serves only as a launchpoint for the far vaster database of resources across the web for which the student has now developed a literacy for engaging.

As this information accumulates, some areas become well traveled. They are the foundational areas of the curriculum and the most popular topics among students and faculty. However, there are also less traveled areas, subjects that are only occasionally covered or reflect interests that are not widely shared among the students. These areas form a mini long tail or more accurately a portal into the long tail effect of the Internet. For example, I teach an upper-division course on contemporary poetics once every two years or so. It's a small course to begin with, and maybe one student in the course becomes especially interested in the language poets. She posts about some of the poets involved, discusses her impressions of their poems and manifestos, and provides links to various sites (e.g. Ron Silliman has a blog). Her work in itself may not amount to much, but it provides a starting point, a way into this world. While some of the links will degrade over time, the student's work retains value because of the long tail, because the information remains accessible for the small number of students over time who will take interest in it. Clearly these qualities scale very well. In fact, they become more pronounced with an increase in users (witness Wikipedia). While three tenured faculty, a few instructors, a couple dozen majors, and a hundred other students each semester will certainly make something of value this way, one imagines the value increases substantially as numbers increase. Certainly one could imagine a "national" or even international disciplinary wiki, but there is also something to be said for the value of local knowledge and practices, the record of a particular community, especially when that record integrates seamlessly into the larger network.

Throughout this integration of technology into the curriculum, it is important that courses not only use the technology but also foster rhetorical awareness of its functionality and design. Some applications will work better than others and certainly different students will have different reactions. As the students shape their own user experiences and rely upon the larger network of information produced through the curriculum, they will have an opportunity to develop their design sensibilities and apply them to this environment as well as other areas of their lives. The Web 2.0 environment I am describing is not monolithic. It is not an all-in-one system like Blackboard or WebCT. Instead it is a constellation of applications produced by different companies and connected by common standards (e.g., RSS) and shared APIs. As a result, students get to encounter a variety of design approaches and give thought to the different ways these applications can be interconnected. For example, students need to consider how to bring together various streams of information on their blogs. Such sites are not only for their personal use but also are a means for producing an online identity for themselves. On this level, students need to consider the design and arrangement of information as a user experience.

Overall, Web 2.0 technologies offer a powerful means to produce, distribute, and organize the knowledge of a disciplinary community. For Professional Writing, providing students with experience with new technologies is valuable in itself, as those skills prove marketable in the workplace. More importantly, however, this curricular shift leads to new classroom practices and epistemologies that will prepare students for the professional tasks of the emerging economy. In working with folksonomic tagging, students learn to recognize that their education does not fall neatly into discrete categories but rather is distributed across an open space where it is subject to *post-hoc* organization. Technical writing and fiction writing may represent different segments in the curriculum, but that does not mean that technical writers cannot benefit from understanding narration or characterization or from learning to create a sense of empathy. Likewise a fiction writer can come to see that his storytelling skills are not only applicable to writing short stories or novels but intersect a range of possible careers. By shifting the entire curriculum in this direction, we move away from the curious genre of the classroom academic, researched essay with its vague purpose and audience of one. Students continue to do research and continue to make arguments. They simply do so now in a shared communal space. Such a practice may not be appropriate for first-year composition students, who may have serious struggles with writing and may not want their work shared publicly, but for Professional Writing majors seeking careers as writers, the Web 2.0 environment provides a context where they can put their entire repertoire of skills to work.

THE FACULTY DEVELOPMENT BOTTLENECK

Of course the caveat here is that one must have faculty with the necessary skills and the commitment to keep up with emerging technologies. Doing so means not only keeping abreast of new developments and learning how to use them, it also requires thinking about their uses in the classroom and their integration into one's particular courses. In short, while faculty from a generation ago speak about working to "perfect" their courses (so that they could then be replicated year after year), here we face the prospect of regularly retooling. This demand changes expectations regarding faculty training and curriculum development. Even a few years ago we might have said that a professor could choose whether or not to learn how to use a course-management system like WebCT. A college might provide some incentive for faculty to learn the technology, but the premise was that faculty who did not learn new technologies could continue to teach their courses by traditional means as effectively as they had in the past. Now, however, by not integrating technology into courses, faculty fall short of addressing the ways in which emerging technologies are shifting the production of knowledge across the culture and in every discipline (especially as those disciplines function at the level our students will employ them as professionals with undergraduate degrees). In other words, the integration of technology is becoming an increasingly necessary element of higher education, and few colleges are likely prepared to face such a necessity.

Fortunately on the scale of individual professional writing programs, the problem is much more manageable, though certainly the characteristics of the challenges are largely shaped by local conditions. However, since professional writing programs have commonly formed in response to the demands of the workplace, most include at least one faculty member with a degree of specialization in new media. In our situation, our program and faculty are relatively small. As such it's possible for me to support my colleagues, to call their attention to new developments, and to troubleshoot with them or their students in our computer lab. We can easily meet and discuss ideas for our courses, where infusing technology is only one of many issues we might raise. Of course the college also provides technical support and training, but the advantage of our working together is that I can discuss technology with them from a shared disciplinary perspective and with a common understanding of our students and our program's goals. While this works on a small scale, there is no way I could perform this same function for the dozen other faculty who teach literature or English education in my department, even if there was an interest on their part in my doing so. That said, there are several qualities of our approach that reflect a more general strategy for professional development and the infusion of technology into curriculum.

Again, here it is useful to turn to the broader professional discussion regarding the integration of emerging technologies into the workplace, academic or otherwise. The strategies for doing so largely reflect the central tenets of Web 2.0 development, such as trusting users, which I mentioned earlier. Within an institutional hierarchy there are essentially two modes for implementing any kind of policy change: top-down and bottom-up. Top-down approaches are common in corporate structures and occur in certain contexts within academia (perhaps increasingly so). However, faculty are generally resistant to administrative decrees, *especially* in the area of curriculum. Nevertheless, the model is not untypical in relation to implementing technologies on campus. In the past, implementing new technologies has required significant capital investment to purchase new software and hardware, to update existing networks and machines to ensure compatibility, and to train and/or hire support staff. In this context, colleges have sought to regulate the use of new technologies. In large part this comes out of an underlying mistrust of faculty and students. In terms of faculty, there is a (perhaps not inaccurate) perception that professors need to be trained and constrained otherwise they might become frustrated with the learning curve of new technologies or somehow "break" them, causing support nightmares. The mistrust of students is even greater; students might use emerging technologies for any number of activities (Napster, for example). So when my college implemented WebCT, it required faculty to go through a multi-day training course (even those of us who had already used a CMS at other institutions). Even then, faculty could not create their own courses or add student users to their existing courses. Our ability to make use of the system was kept at the absolute minimum necessary to run a course online. Naturally student users had even less control.

Of course the purpose of an application like WebCT is course *management*: the management of individual courses by faculty, *and* the management of the collective course offerings on an administrative level. It's a piece of software that embodies the top-down thinking of a pre-Web 2.0 environment and knowledge industry. On the other hand, Web 2.0 technologies are largely native to a bottom-up approach. Unlike their predecessors, they are inexpensive. In fact, many are free to users or at least offer a free level of membership. They are designed to be "light" and compatible with standard, if up to date, web browsers. As such they require little capital outlay from an institution in order to be implemented. As I've already mentioned, the primary challenge and expense here is faculty development. Fortunately, the bottom-up approach offers a different model of development that rests upon trusting users to develop powerful uses for new technologies. This approach works equally well for getting both faculty and students involved.

As Suw Charman, a social software consultant and author of the well-regarded blog *Strange Attractor*, describes, a grassroots approach to technology implementation begins with identifying key user groups and specific individuals within those groups. Focusing on my primary concern, a professional writing program, the key user groups, at least for us, would be faculty, instructors, and students. I separate tenured faculty from instructors because the faculty have a wider range of responsibilities in relation to the program. Beginning on the group level, Charman lists a few important questions: "What needs do these people share? What are their day-to-day aims? What projects are they working on together? What information flows between them, and how?" Without going into specific responses here, clearly the traditional sharing of aims, projects, and information takes place course-centrically between students and the course instructor and perhaps among the students as well. Information flow among instructors and faculty is more idiosyncratic, except in the case of formal, departmental class observations and personnel review. As I have already suggested, the idea here is to shift these relations and practices, but Charman's point is that one must begin by responding to users' existing practices and needs.

Once these questions have been answered, one needs to identify key individuals among the groups who are well-connected and potentially interested in the technology. Principally the idea is that one would take advantage of existing social networks (which are notably different from institutional, hierarchical relationships). Convincing the right individual to adopt a technology will lead to others following his or her lead, even among largely independent-minded faculty. Again, Charman posts some questions that are relevant here: "What specific problems does social software solve? What are the benefits for this person? How can the software be simply integrated into their existing working processes? How does social software lower their work load, or the cognitive load associated with doing specific tasks?" On the student level, one might identify a student who is well-known and perhaps popular among her peers, someone with an ability to motivate and convince others. In our case, that individual might be the editor of our literary magazine or the president of our program's student group, the Cortland Writers' Association. These students have problems communicating with other students in the program and organizing various activities that social software might solve. Using social software allows them to separate the business of these activities from their personal email or mobile phones and reduces their responsibility for keeping contact information. For example, when editing the literary magazine, one of the more onerous tasks is keeping track of the submissions, which submissions have been rejected or accepted, which need to be evaluated, which need to be edited, and so on. A project management application, like 37signal's Basecamp, would greatly reduce both work load and cognitive load.

Charman's idea is to turn these individuals into "evangelists" for the application's use and eventually into trainers. This might work out a little differently on the student level. However, to continue my example, if the student editor of the literary magazine started using Basecamp and convinced other students to use it as well, they would quickly discover whether or not the application did make their lives easier. If so, they might start experimenting with using it for other purposes. With this bottom-up approach, it is important at this point that someone from the top does not come down and restrict these uses. If the students use the application to plan a party, or for other non-academic social purposes, this should be encouraged or at least not discouraged. Simultaneously, one would seek to foster adoption among instructors and faculty, perhaps pointing out how project management software might help with managing longer, multi-assignment, projects in the classroom. It might also benefit full-time faculty with other service obligations like tracking curriculum development or assessment. Thus, when Basecamp is introduced into a classroom a number of the students might already know the application and have positive associations with it. These students can help the instructor by supporting the other students. Likewise, more experienced users among the instructors and faculty can serve as an informal resource for their colleagues.

Of course, such work can only go so far without some top-down support. Student use must be supported by instructors and faculty, both by incorporating the application into coursework and encouraging non-academic uses the students discover. Likewise, instructors and faculty need support from administrators. They need the work they do to learn new technologies and incorporate them into teaching to be recognized and rewarded in their personnel evaluations. They need the service they provide as trainers and as support resources to be factored into their workload. As Charman point out, it's also important that the adoption of a new technology be reflected in the institutional hierarchy and its daily workings. It's difficult to encourage students to use electronic communication if the teacher always responds on paper. Likewise, faculty will find it difficult to value a new mode of information sharing, like a wiki for example, if the department chair keeps sending out paper memos or even emails. Furthermore, these initial users-turned-trainers will eventually need assistance from the college's formal technical support personnel. As adoption of an application grows, the college's support professionals will be more generally available to answer questions from students in their dorm rooms and faculty at home. In other words, at some point, a bottom-up approach will stall out unless it is affirmed by some clear signal coming down the institutional structure. Ross Mayfield refers to this coming together of the bottom and top as the creation of a "middlespace."

While faculty development and the challenges of adopting new technologies may seem tangential to the curricular goals of Professional Writing, these issues actually fit in quite well. Indeed, as Jon Udell's articulation of screencasting describes, one of the growing needs in his industry is for writers who can not only perform the "technical writer" task of describing an application's various functions but also can succeed at the more creative challenge of reaching potential users and helping them to envision how a new application might fit into their social practices. This rhetorical goal, put more generally, is part of the skill set Daniel Pink identifies for workers in the post-knowledge economy: the ability to empathize with users and design powerful user experiences. The task of fostering the adoption of emerging technologies within an undergraduate program becomes one instance of the rhetorical work graduates will later find themselves doing.

MOVING FORWARD

The Web 2.0 technologies I've been describing here are the current emerging technologies. Indeed blogs and wikis have been around for several years. EDUCAUSE and the New Media Consortium produce an annual "Horizon Report," which identifies emerging technologies. In that report, they identify the technologies I have been discussing as those that are currently being implemented or will be implemented in higher education in the next year or so. Beyond that, they point to the growing functionality of mobile phones and the influence of educational gaming in the next two to three years; and the implementation of what they term "augmented reality," "enhanced visualization," and "context-aware environments and devices" in the next three to five years. Most of these latter technologies are already seeing use in the sciences, medicine, and engineering. The role they will play in Professional Writing curricula is obviously uncertain. This does not mean that current technologies will fade away, though it is likely they will change. It is not too difficult to imagine how a podcast or wiki might evolve to work with a mobile phone or in a "context-aware" environment. And with a little imagination, one can see how these folksonomic information structures might operate using a game-like interface or via three-dimensional, virtual modeling. Regardless, inasmuch as these emerging technologies compose, design, communicate, and organize information for user experiences, they will have rhetorical elements that our discipline will be able to address. And insomuch as these technologies become part of the marketplace, there will be careers for professional writers who can use these technologies, evaluate them for others, and support their use.

In short, while the current wave of technologies from blogs and wikis to social bookmarks clearly has an intimate relationship to writers, we should expect that evaluating and adopting new technologies into Professional Writing will be a regular feature of our careers for the foreseeable future. As such, it is imperative when designing a new curriculum that one attempt to incorporate structures that will accommodate such a practice.

NOTES

[1] A screencast is generally a video capture of a computer desktop complemented by a voiceover. In a screencast a particular application is demonstrated on the video capture of the desktop as the user explains the various steps he or she is taking. A screencast might also be a Powerpoint or Flash-driven set of slides, again accompanied by audio. Like audio and video podcasts, screencasts can be published in a blog format or distributed via RSS.

[2] For more on Web 2.0 read Tim O'Reilly's "What is Web 2.0?" O'Reilly provides an excellent analysis of the primary features of Web 2.0 technologies. Christopher Allen provides a useful chronology of the concept of Social Software, and Clay Shirky's "Social Software and the Politics of Groups" provides further insight into the concept (Shirky is generally credited with conceptualizing contemporary Social Software). Both Web 2.0 and Social Software are elements in a lengthier objective that is termed the "Semantic Web." This concept was coined by Tim Berners-Lee (who also developed the http protocol that essentially created the web in the late eighties).

[3] What are API, AJAX, and RSS?

API stands for "application programming interface." It is the interface that allows one program to request information from another. For example, using the Wikipedia API, someone could develop a web-based application that called up and displayed information from the encyclopedia. While some APIs are closely-guarded and proprietary, many Web 2.0 applications openly share their APIs with the belief that the more other sites make use of their service the more valuable their service will become.

AJAX stands for "Asynchronous Javascript And XML" and references a programming strategy for the web. Without going into great technical detail, AJAX allows for a more seamless experience of the web, where the browser needs to make fewer requests of the web server. Google Maps (http://maps.google.

com) is an example of AJAX at work. On a site like Google Maps, users can drag across maps, zoom in and out, and switch map views from road maps to satellite pictures without the page having to reload.

RSS stands for "Really Simple Syndication" and is an XML file format that allows for the distribution of the content of websites. When one subscribes to a podcast or to a blog and receives that information through a blog aggregator either online (like Bloglines) or on the desktop (like podcasts in iTunes), one is connecting to an RSS file (also called a "feed") that is generated automatically by blogs when bloggers publish their posts. RSS is not limited to blogging however. Any dynamic or regularly updated website or database could generate an RSS feed. Wikipedia provides more detail on each of these terms.

[4] The principle of the long tail identifies the opening of new markets for products outside the mainstream. For example, a local department store can only carry a limited variety of CDs. However beyond these most popular products there exists a long tail of products that appeal to smaller groups of people. While a brick-and-mortar store must cater to a local population, through the web, companies can appeal to a smaller, vertical market.

[5] One of the many examples of this warning against academics is the pseudonymous "Ivan Tribble," who has published several screeds against blogging, including "Bloggers Need Not Apply." Oddly, there seems to be little awareness on "Tribble's" part of how much his own anonymous column resembles the negative characteristics "he" sees in others blogs.

[6] Folksonomy (as opposed to taxonomy) is the practice of tagging websites (and other media) with one's own keywords and then sharing keywords with others. This makes use of a primary advantage of electronic over print information. Books in a library can be organized in only one way (i.e. a book can be in only one place), electronic data can be organized differently by each user. Library systems are *ad hoc;* the system predates the books that become organized. Folksonomic systems are *post hoc;* they describe media after its publication. For more on folksonomy, read Clay Shirky's "Ontology is Overrated: Categories, Links, and Tags."

WORKS CITED

Allen, Christopher. "Tracing the Evolution of Social Software." 2004: 11 March 2006. http://www.lifewithalacrity.com/2004/10/tracing_the_evo.html.

Anderson, Chris. "The Long Tail." *Wired.* 12.10 (December 2004): 5 March 2006. http://www.wired.com/wired/archive/12.10/tail.html.

Berners-Lee, Tim et al. "The Semantic Web." *Scientific American* (May 2001): 10 March 2006. http://www.sciam.com/article.cfm?articleID=00048144-10D2-1C70-84A9809EC588EF21.

Charman, Suw. "An Adoption Strategy for Social Software in Enterprise." *Strange Attractor*. 5 March 2006. http://strange.corante.com/archives/2006/03/05/an_adoption_strategy_for_social_software_in_enterprise.php.

EDUCAUSE and the New Media Consortium. *The Horizon Report*. 2006: 3 March 2006. http://www.nmc.org/horizon/.

Gabriel, Richard. "Master of Fine Arts in Software." 2001: 12 March 2006. http://dreamsongs.com/MFASoftware.html.

Mayfield, Ross. "Middlespace." *Many-to-Many* 27 October 2004. 5 March 2006. http://many.corante.com/archives/2004/10/27/middlespace.php.

O'Reilly, Tim. "What is Web 2.0?" 30 September 2005. 5 March 2006. http://www.oreillynet.com/pub/a/oreilly/tim/news/2005/09/30/what-is-web-20.html.

Pink, Daniel. "Revenge of the Right Brain." *Wired* 13.2 (February 2005): 5 March 2006. http://www.wired.com/wired/archive/13.02/brain.html.

Pink, Daniel. *A Whole New Mind: Moving from the Information Age to the Conceptual Age*. New York; Riverhead, 2005.

Shirky, Clay. "Ontology is Overrated: Categories, Links, and Tags." May 2005. 15 March 2006. http://shirky.com/writings/ontology_overrated.html.

Shirky, Clay. "Social Software and the Politics of Groups." 2003. 7 March 2006. http://shirky.com/writings/group_politics.html.

Silliman, Ron. *Silliman's Blog*. 2006. 10 March 2006. (http://ronsilliman.blogspot.com/).

Tribble, Ivan. "Bloggers Need Not Apply." *Chronicle of Higher Education*. July 2005. 9 March 2006. http://chronicle.com/jobs/2005/07/2005070801c.htm.

Udell, Jon. "The New Freshman Comp." *O'Reilly Network* 22 April 2005. 10 March 2006. http://www.oreillynet.com/pub/a/network/ 2005/04/22/primetime.html.

POST-SCRIPTS BY VETERAN PROGRAM DESIGNERS

15 A *Techné* for Citizens: Service-Learning, Conversation, and Community

James Dubinsky

"It is a natural mistake to think that reverence belongs to religion. It belongs, rather, to community."

— Woodruff, *Reverence*, 5

In *The Year of Magical Thinking,* Joan Didion attempts to understand the grief she experienced and explain the emotional and practical tasks she faced in a year that began with the sudden death of her husband. He suffered a massive heart attack, hours after visiting their only daughter who was lying, near death, in a nearby hospital. To situate us and begin the narrative, she describes these events to illustrate how "Life changes fast. / Life changes in an instant" (3). And, despite tremendous grief and bewilderment, after a year of reflection, she comes to a conclusion that to survive "you ha[ve] to feel the swell change. You ha[ve] to go with the change" (227).

While I am not writing about grief, which was Didion's catalyst, I am writing about change and the ability to think "magically" to deal with stressful, difficult, and unexpected issues. As I look back on my ten years at Virginia Tech as a program builder/administrator, I am convinced that such an ability is necessary for almost all program builders in our field. In 1998, I arrived at Virginia Tech, a newly "minted" PhD, who faced a difficult task on top of the standard "research/publish, teach, and serve": I was asked to build a professional writing program in a traditional department of English; revise two service courses in business and technical writing, one of which was under tremendous pressure due to some unusual (erratic) teaching; and lay the foundation for future graduate study. Much like me, most recent PhD graduates who take administrative positions in our field come from programs that understand and value technical and professional writing. And many, if not most, get hired by English departments that may not value and probably do not understand it. They, like Didion and me, discover that life changes fast, and responding and adapting to that change requires something akin to "magical thinking."

I wish I could promise that what follows is the formula for such magic. However, I doubt it is. Rather, it is a story about my first year and the five that followed as I attempted to adapt to these changes. And it is a story about what Paul Woodruff, in his fascinating study of the virtue of reverence, calls "the paradox of respect" (197). Woodruff explains that reverence is "the capacity for certain feelings where they are due" and one of those is respect. However, knowing when they are due is not so easy because "respect comes in three degrees of thickness: too thick, too thin, and just right." The respect that has the degree of thickness that is "just right" is the type that "flows from reverence"; it involves "a felt recognition of a connection growing out of common practices" (198). Woodruff's work is relevant on several levels. First, if hired by a traditional department of English, more than likely you will experience such a paradox. Your new colleagues, because they hired you from among many candidates, will respect your credentials and potential. However, they will also not know enough about your work to make a strong connection to you intellectually or even perhaps emotionally. In addition, there will be, at least among some of your colleagues, a certain amount of trepidation and worry and perhaps lack of respect. Will your work "fit in"? Will you add to or take away from their department's reputation for scholarship? Will the applied focus you offer dilute their focus? Thus, one of your tasks, implicitly or explicitly, is to create or at least encourage a "felt recognition" based on "common practices." Those practices will encompass not only your dedication to your epistemological position but also, and perhaps more importantly, your dedication both to the art or *techne* of teaching and to the *techne* of citizenship, both within the department and in the larger cultures—of the university and beyond.

My hope is that this "story" of our program at Virginia Tech will provide insight and some answers for those engaged in programmatic work. I see it as part of a move in our field to treat curriculum as conversation, an important shift toward ways of knowing that are more explicit, that work to articulate what Polanyi and others call "knowledge-in-action" or "tacit knowledge" (Applebee 11). Such a focus creates a domain where engagement with new texts and issues can lead to discovery and transformation. One such issue is a question that is embedded in the debate between art and science framed by C. P. Snow and manifested in many PTW programs as a fundamental tension between developing "insight or technique, liberal or vocational education, good citizens or good workers" as discussed in the preface to this volume. Rather than resolve this dilemma, my goal has been use this tension productively. As a program designer and leader my critical question, a mission statement of sorts, is simply this: how can I shape a PTW program that will graduate informed, critical citizens who can use their technical expertise for public service?

As I tell this story and address this curricular question, I hope that our curriculum at Virginia Tech will become a text for conversation that has relevance to our field. Our program's emphasis on rhetoric and experiential learning and our focus on principles of reflective practice such as "open mindedness" and "responsibility" (Dewey 177) have enabled us to create a curriculum that is both epistemic and instrumental, one that balances theory and practice, phronesis and praxis. This emphasis on key principles has helped us overcome some of the problems discussed in this volume, such as the issues of naming and other issues such as the politics of identity caused by the tension between liberal and practical arts.

The professional writing curriculum at Virginia Tech is part of a larger English curriculum that gives students more control and experience with technology, as well as opportunities to apply their knowledge and expertise for the benefit of others. The curriculum, as a whole, seeks to define "common practices" emerging from these principles of reflective practice, such as a belief in reflection and assessment, which are both essential components of effective teaching and learning. Currently this belief informs the department's recent adoption of ePortfolios as a strategy to improve teaching and learning. In addition, many faculty in composition studies, creative writing, and in our program use service-learning or client projects. All of us believe in teaching students the value of self-reflection, critical reading and analysis, and a multicultural context, three of six essential learning objectives our faculty have agreed are essential for all English majors. Despite occasional difficulties, this agreement on common practices has led to a form of reverence and respect among the faculty that "does not stop at boundaries" and "overlooks differences of culture" (Woodruff 84). The result, we believe, is a *techne* for citizens" (de Romilly 30) in which students gain qualities Cicero believed were essential to making human social life possible: practical experience, expert knowledge, and a sense of responsibility for private and public life (6).

PROFESSIONAL WRITING: THEN (1998) AND NOW

To understand how our program and department have come to operate with what I believe is a form of reverence, I begin with some contextual/historical information. I am in my tenth year of service to the English Department at Virginia Tech, and during this entire time I have been responsible for the professional writing program, even before a program existed. In 1998, the year I was hired, Virginia Tech, a Carnegie-rated research level I university with a traditional department of English, had no program in technical or professional writing, unlike many of its peers (e.g., Purdue or North Carolina State). Our department had 101 personnel, but only one with a PhD in Rhetoric and Composition or Professional Communication. The department's emphasis was literature, although there were a number of prominent creative writers on the faculty as well. English majors could add a Professional Writing Cluster, but this "cluster" was hollow, consisting of the two service courses (English 3774—Business Writing and English 3764—Technical Writing) and a course in Advanced Composition, which had no defined, consistent content. Equally important to note is a fact that I learned during my interview process that the reputation of these service courses in other colleges, particularly in the College of Business, had been diminishing.[1] There were no graduate courses in professional or technical writing and only one in composition—the required course in pedagogy for the graduate assistants, taught by Dr. Paul Heilker, who then directed the program in Composition.

Despite what I've just outlined, my position was highly vulnerable. During my interview with the dean, I learned that I was "an experiment"—the department's first tenure-track faculty with an advanced degree in Rhetoric and Professional Communication. For the first three years, the dean's office would pay sixty percent of my salary, and the dean told me quite directly that my contract renewal would depend on my success in reviving the credibility of the two "service" courses (Business and Technical Writing), expanding our emphasis in outreach, integrating technology into the curricula more effectively, and creating a program in Professional Writing, one that could be extended into a graduate program.[2] Despite the administrative duties involved, I would teach a full load and would be expected to conduct and publish research. The challenge, needless to say, was daunting.

Ten years later our department has ninety personnel—fewer overall, but a higher percentage of permanent faculty and virtually no adjunct or temporary faculty. More to the point, now eleven of the ninety (or twelve percent of total and over twenty-five percent of tenure track) have PhDs in Rhetoric, Composition, and/or Professional Communication. Several of these eleven are senior hires who have significant administrative roles in and out of the department (e.g., Carolyn Rude is our department chair; Diana George directs our Composition program and Writing Center; Kelly Belanger directs our Center for the Study of Rhetoric in Society; and Shelli Fowler directs a major university initiative for graduate education). Our Professional Writing Program not only exists, but it is one of the three options for English majors. We counted, as of last spring (2008), just over two hundred majors and minors, and our curriculum consists of nine courses at the undergraduate (3000-4000) level in addition to the two, previously mentioned, service courses. Equally significant is the fact that enrollment in those two service courses has more than doubled (eleven sections of English 3774 and twenty-seven of English 3764), and we've quadrupled the number of online sections. We regularly offer graduate courses at the MA level, and we now have a PhD in Rhetoric and Writing.

STARTING FROM SCRATCH

As is evident with the brief overview, much has changed in the years since I arrived. But, like a relative always told me when I faced a very large project: one eats an elephant one bite at a time. So to help understand what has happened, it helps to understand my approach, which was to treat this task like a qualitative research project. In essence, I began with a *needs assessment*.

Before I could consider making any changes, I had to understand what was present. I needed first-hand knowledge of our current service courses and their impact on our majors and on the university at large, because so many departments required their students to take them. To address that lack of knowledge, I began by teaching both courses. In addition, I surveyed the faculty who were and had been teaching them and examined their syllabi. Finally, to understand why people, particularly those in other departments, were dissatisfied with these courses, I visited our "client" departments and colleges, talking with their curriculum committees and surveying their faculty.

What I learned, particularly from the School of Business faculty, who six years before had "delisted" English 3774 as a required course, was useful. I learned that many departments would prefer that their students take a course dedicated to writing taught by qualified faculty, but they had been unsatisfied with the instruction previously. Our department's credibility was damaged, and my interviews with current English faculty confirmed that they were aware of the problem.

Most important, I gathered perceptions that other departments had of our department's writing programs, to include first-year composition, on whose curriculum committee I sat. I learned, in detail, what other departments hoped we would teach their students. I learned where our writing courses fit into their curricula. I also learned a lot about the ways in which faculty across the university saw writing in general, whether or not they felt competent to integrate it into their own curricula, and why they believed students needed to learn to communicate. I learned, in effect, what they envisioned or knew about writing and writing instruction. Bringing this information back to the department proved helpful in many ways, not the least of which was community building, which was perhaps the first step toward "common practices."

During conversations with the directors of our Center for Excellence in Undergraduate Teaching, the Office of Educational Technologies, and several associate provosts, I learned about the directions the university was moving as it revisioned its strategic plan and about several major curricular initiatives concerning undergraduate education. I learned, for instance, that the university intended to focus more intently on enhancing its status as a research institution while reviving its land-grant mission and outreach/engagement. One of the foci to achieving both goals would be to use technology to increase access and build bridges to communities. The Center for Human-Computer Interaction (HCI) was already involved in a number of such projects with local schools, and I became an affiliated faculty member.

BUILDING BRIDGES & FINDING COMMON GROUND: LONG-TERM STRATEGIES

My data gathering and subsequent reflection led to my taking some immediate steps, with an eye toward developing long-term goals. My immediate steps focused on my own pedagogy to see if I might develop a model for others to follow. To that end, I enlisted the assistance of the Service-Learning Center to introduce me to nonprofit agencies and to help me learn more about our community's needs. I believed that service-learning pedagogy was a route to create engaged citizens, a topic I subsequently wrote about ("Service-learning"). This strategy fit in well with the university's long-term goal of revitalizing its land-grant and outreach missions. And it enabled me to give students hands-on, experiential learning opportunities, ones that, if successful, would also build both their resumes and excitement, which I hoped would help to enhance our department's credibility.

Longer term strategies were equally, if not more important. I was in a department that was both welcoming and a bit wary of my presence. Historically a literature department, the faculty recognized that the number of majors had dropped considerably (about three hundred when I arrived). Still, there were strong opinions about service, service courses, and being considered a service department that I had to overcome. As I've written about elsewhere, my strategy was to build bridges and demonstrate that, while professional and technical writing were applied disciplines, they were not vocational ("Status of Service"). In addition, our discipline produced knowledge and often relied on some of the same research methods used by those in literature. One of my essential tactics in that strategy involved a capstone course, rooted in rhetorical analysis, that would help students recognize the impact that texts have on public policy. Thus, at the end of their program, after working on project-based, experiential learning courses, professional writing students would step back and analyze the impact that the kinds of texts they had been creating could have in a variety of contexts. Rhetoric and its historical connection to teaching and to the roots of all literature departments became my bridge (Thelin, Beale, Rudolph, Murphy).

The capstone course, ultimately entitled Issues in Professional and Public Discourse, became the senior seminar for professional writing students. To have this course qualify for the seminar status, which did not occur for several years,[3] I had to demonstrate to my colleagues on the undergraduate curriculum committee how it met the pre-defined criteria for senior seminars designed for literature majors. These criteria centered around two key issues: research and analysis of central texts. In such a capstone course, students would have to develop "the ability to demonstrate knowledge and understanding of the discursive, social, historical, biographical, or cultural contexts out of which or against which literature is written" (Rubric). The challenge was to create an argument that substituted nonfiction policy texts for literature. Doing so did not prove to be nearly as difficult as one might expect with the more recent interests of some faculty in popular culture, in historiography, and in social theory.

While I was gathering data, teaching, and rethinking the service courses, I was also meeting and becoming familiar with the strong core of faculty who were dedicated to teaching writing – both composition and technical writing. I recruited a few interested instructors, two with PhDs, to be on what started out as a brownbag discussion group on professional writing (PW). This group met once every three weeks. I also received an outreach grant from our college, which enabled me to hold a one-day colloquy (Bringing Business to Business Writing) and put together an initial website that we used as a basis for our current program home. At the colloquy, I met some people who were then working in industry as tech writers, and encouraged them to come back to teach, outlining the potential of the new program. One, Marie Paretti, did, and, with her exceptional background, was hired. Thus at the end of year one, a core group of faculty were gathered, and the roots were growing.

YEAR 2: WRITING PROPOSALS AND FUNDING CHANGE

In the fall of my second year, the department appointed me head of a PW Task Force and for the next two years, our core group of faculty studied key programs and curricula across the country, and I networked with other program developers, using CPTSC and ATTW as forums. With some curricular data to use for support, we then asked for money to find out whether or not the university truly was behind this "experiment." We submitted grant proposals for course development, using most of the money to buy down instructor loads (from 4/4 to 3/3), which would give them time to assist me with course development (to include creating materials for an online course).[4]

In those proposals, our arguments, at first, were quite pragmatic. We began by examining the university's own documents, such as the College of Arts and Sciences' Annual Report. In that document, we discovered that the college asked the departments to meet Objective 5.2, which had as a goal "to assure ourselves and our publics that we prepare students appropriately to become professionals" (1). We argued that the ad hoc group become more substantive and embedded in the fabric of the department and college. Doing so would give us credibility and visibility.

We also looked at institutional research, where a wealth of survey data resides. By examining alumni surveys, we learned that fifty-eight percent of recent graduates indicated that writing is of "great" or "critical" importance, and eighty-three percent indicated that writing is of at least "some" importance in nearly every profession surveyed. In these surveys, graduates also focused on the need to develop problem-solving and oral communication skills.

We then connected the dots, making linkages between these skills and the university's increasing emphasis on information technology, which was so evident in all of its recent communications, particularly its two magazines focusing on research and alumni relations. Using data such as job lists in *Money* magazine in which technical writing was listed as one of the ten hottest professions,[5] we explained that technical/professional writers work in fields as diverse as computer software documentation, engineering, science, and medicine. They also work as WWW designers, information and media specialists for multimedia companies, and in business corporations.

We argued that courses in the proposed professional writing track would address several of the university's strategic concerns. The skill sets that students would develop and the practical experience they would obtain would help them better achieve their career goals and prepare them to be life-long learners. Equally, if not more important, we argued that these same skill sets, when applied using service-learning or client project pedagogy would help them mature into professional citizens of the world. They would become, in Cicero's terms, ideal orators.

As I said, our initial focus was on pragmatic concerns: teaching students to write clear, coherent prose; adapting their skills to meet the demands of changing technologies; gathering and interpreting data; and planning and managing projects. We wanted to assist them to develop transferable skills by providing hands-on opportunities with many of the leading software packages (word processing, spreadsheet, graphic arts, WWW design, telecommunication, etc.).

We also drew from the alumni surveys to argue that a program in professional writing was long overdue. Several alumni had commented that a "concentration or degree [in professional writing] would be a good addition," and a recent outside evaluator, surprised at our current configuration, had said that "Virginia Tech would seem to be an ideal place for . . . valuable training in technical matters and clear prose." We buttressed this claim with an argument from example, explaining that many of our peer institutions (e.g., Purdue) had recognized the need for such programs long ago and had thriving programs that were continuing to grow.

Finally, we tackled one of the most difficult arguments—that of professional and technical writing being seen as too applied or worse, too vocational. We reasoned that if the goals were to insure that students have the skills necessary for success and to prepare students to be competitive for the jobs described above, the university needed to support courses that develop professional competencies. Recognizing that for many years, such courses have met with resistance,[6] we argued that our potential program provided an opportunity to reverse that trend.

But pragmatism and logic only go so far. We also wanted to appeal to the hearts of many of the university leaders, who were starting to resurrect key initiatives tied to Virginia Tech's historic, land-grant mission. Because I had already experienced success with the service-learning projects I'd tried ("Service-Learning"), we chose to highlight the opportunities to work with clients in the nonprofit sector as part of a coordinated service-learning strand that we intended to thread through the program. I began using this pedagogical strategy in the service courses (Business and Technical Writing), then I integrated it into every pilot course I taught, as did my colleagues. I began to work closely with the Service-Learning Center, and I was fortunate to win two university awards, which elevated the work's value in the eyes of my colleagues and led to further pedagogical discussions, both formal and informal. Several new hires, such as my current assistant director who is also a director of a non-profit organization, applied because of the emphasis on service-learning that we had. Students were receiving additional internship and co-op opportunities, and we received positive affirmation from the college. The service-learning component provided students with valuable experience as they applied concepts learned in academic contexts to real-world need. More to the point of program building, it offered needed credibility and visibility. As a result, our appeals using student and community partner testimony were very effective. These appeals, I learned later, sealed the deal. And we were funded.

The funding enabled us to develop and pilot several courses such as the Rhetoric of Disaster and Discovery (a predecessor of the current capstone course mentioned earlier). We also created an outline for a curriculum (see Figure 1), which led to five course proposals being submitted for approval at the university level in the spring of 2000.

YEAR 3: CREDIBILITY AND FACULTY HIRES

In the third year, given we had a curriculum designed and approved and had already submitted six courses to the university, my department chair appointed me director of a "program." However, we still had a long road ahead, as only I could teach the 4000-level courses, and the entire undergraduate curriculum was being revised to adjust to/accommodate this shift of resources toward writing. With my colleagues, we:

- Continued developing & piloting courses
- Submitted course proposals for university approval
- Changed the department's governance and administrative structure to create a Professional Writing Committee
- Requested that one of the instructors with a PhD be appointed assistant director with a one-course relief (down to 4/3)
- Requested English 3764 (Technical Writing) be designated as a Writing Intensive course
- Developed our program website
- Argued for the hire of another assistant professor

All these goals were achieved, and the dean, who had been skeptical three years prior, approved two hires: Eva Brumberger and Jim Collier. What made this decision so important, in terms of the longer-term strategy, was that I was not the one to actually ask for two hires; a senior literature faculty member with an endowed chair did the asking. As a key faculty member on the personnel committee and an active participant in curricular issues, Dr. Ernie Sullivan made the case for these hires after I had presented a status report along with the proposed curriculum to the dean. Having the support of the literature faculty, who had lost a number of positions recently due to retirement, had a powerful impact on the dean. It spoke to culture change in the department and a growing sense that Professional Writing might actually be a valuable addition.

We chose Eva and Jim on the basis of their fields of expertise (Eva's background in rhetoric, technical communication, and composition; and Jim's

PhD in Science and Technology studies). Part of our vision included creating a team who could talk with and serve the various departments/colleges and meet their and their students' needs (looking to attract a variety of students to the minor and hoping to work with the Institutes/Centers springing up across campus in Biotechnology & Leadership for instance): I would handle business /

Courses in the Professional Writing Option
Courses listed with asterisks are required

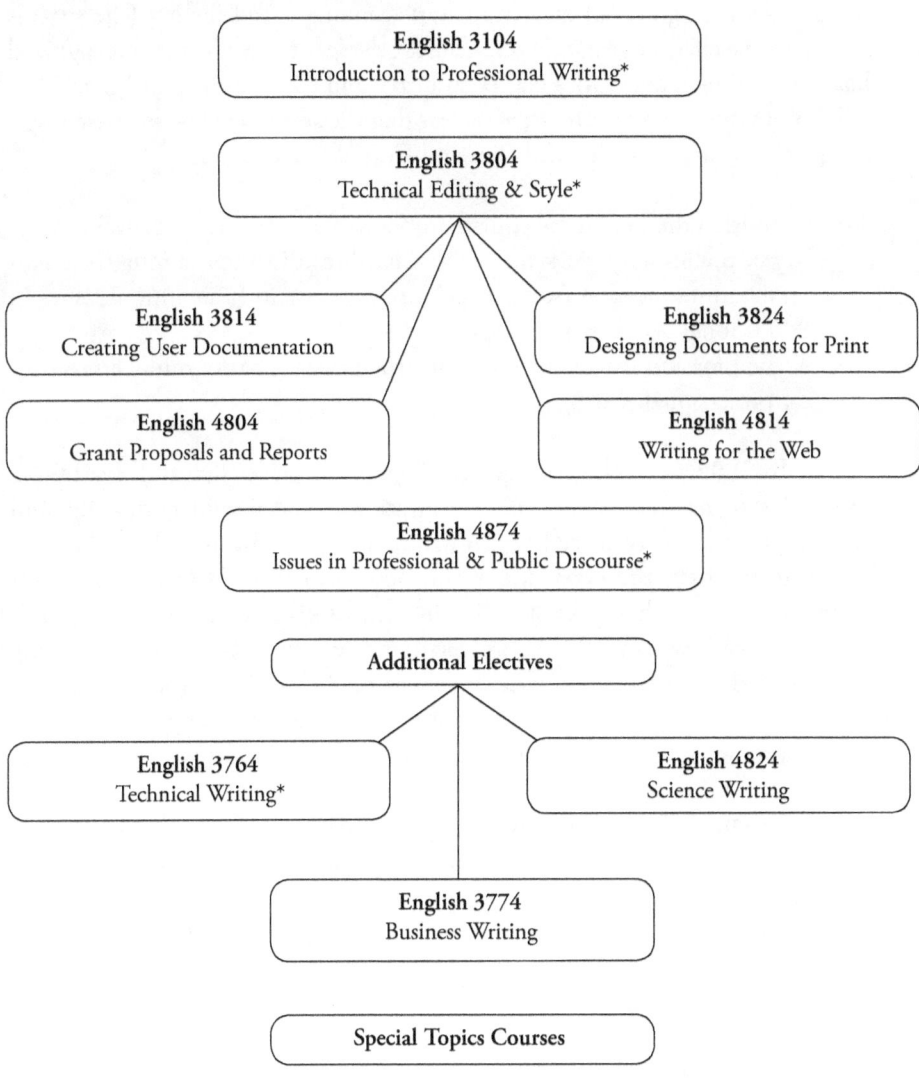

FIGURE 1: DRAFT OF CURRICULUM CA. 2000

educational leadership / outreach; Marie would be excellent for engineering; Jim would handle the natural sciences and philosophy; and Eva would focus on graphic arts and computer science and be a general support for all.

YEARS 4-6: ACHIEVING SUSTAINABILITY

As year four began, with Jim and Eva arriving, we had a foundation and enough of a core group of faculty to start thinking ahead further (the success convinced the chair to give me a one-course release—the department recognized that my administrative duties had been heavy and consuming, and I needed to spend more time on research—it was clear that the admin load wasn't lessening). During this year (2001-02), with Jim and Eva's assistance, we:

- Submitted the final three course proposals for university approval
- Began preliminary work on a potential thread in International/Intercultural communication (we just had a new course in Intercultural Issues in Professional Writing approved)
- Argued for and were able to hire instructors with backgrounds in technical/professional writing.

In 2002, with the core of the program in place, Jim and Eva taking root and starting to establish their reputations—at VT and nationally, and students coming (forty-three English majors interested in the PW option and over forty minors), the department and new dean listened favorably to our argument about further growth—this time in the graduate realm (e.g., a PhD program since it would help the university achieve its strategic goals). Hiring Carolyn Rude (Texas Tech—past president of ATTW) was the result, with our argument being that we could never even hope for a PhD program unless we had a senior scholar / administrator to help guide us, as well as to attract students and other faculty.

Sustainability requires what Carolyn has called a "critical mass" of faculty. While we had hired successfully, maintaining a program means having the junior faculty achieve institutional acceptance, through the tenure and promotion process. Because Paul had been tenured in Composition, we had some hope that the university would value those of us with backgrounds in Rhetoric and accept the kinds of work we did. And, in that critical sixth year, I was fortunate enough to be tenured. Since then, Jim and Eva have been as well, and we've hired additional faculty, enough to grow the program and sustain it.

SERVICE-LEARNING AND THE SCHOLARSHIP OF ENGAGEMENT

I am a firm believer in the work of Ernest Boyer and his argument that there are different kinds of scholarship worth considering. My argument for tenure was predicated, in part, on Boyer's work, and, as such, was not much different than the arguments we had made years before for grant money. I argued, using my dissertation as a starting place, that research, teaching, service, and administrative work are all of a piece, a kind of möbius strip of theory and practice. I focused on reinvigorating the rhetorical concept of *technê* in the field of professional communication, a concept that includes both *art* and *craft* and represents a practical wisdom directed toward some end, a field in which one of the central issues is the study of language use in the public forum in order to prepare students to succeed in that forum as practitioners and citizens. In particular, my emphasis was on *technê's* social, ethical, and rational richness, and its importance to the teaching of writing at the undergraduate and graduate levels due to its connection to civic engagement. Along with my interests in critical thinking and collaborative learning, this focus on technê was significant in my scholarly work and my efforts to build a professional writing program at Virginia Tech.

My goal was to place professional communication within the discipline of English Studies, by focusing on its humanistic elements, while distinguishing it from other fields in the discipline—such as literature—by raising up its connection to service and the practical applications involved that often require research in workplace settings. I argued for process, for a broad understanding not only of writing, but also of the contexts of writing and its impact on people, explaining that much of the research is conducted either in the classroom (the site of learning) or in the workplace (the site of practice). Thus, my argument was that our field requires an interdisciplinary perspective; we work to understand issues of problem solving and critical thinking, usability, document design, cognition, as well as standard issues of grammar and style. We have to be conversant with the latest developments in technology because we teach students to use a variety of media through which meaning is made. And our research methods range from historical and rhetorical to quantitative and qualitative. Like many of our literary colleagues, we study texts, but we also study the use and production of those texts; thus, we are often ethnographers. Finally, because the work is focused on process and because one of the sites of research is also the classroom, we are reflective practitioners who understand the art or *technê* of teaching.

I advocated a user-centered, reflective stance ("Reflective Practitioners"), linking classical rhetorical theory to teacher preparation and the concept of experiential learning. I argued for the importance of both a particular stance—that of a reflective practitioner—and a method of teacher preparation at the graduate level. My emphasis on the concept of civic engagement emerged from a combination of my historical research into the Aristotelian notion of *technê* and the emphasis the field places on practical wisdom. My argument about *technê* came, in part, from making a case for service-learning as a pedagogical strategy that helps students become more reflective, enables them to make both practical and ethical judgments while acting for the public good, and gives them opportunities to apply the key concepts, strategies, and skills they learn ("Service-learning"). Service-learning and client projects served to bridge the gap between practical courses in the curriculum, which are linked to a market-economy, and the ideal of public service, an ideal central to Virginia Tech's culture.

In our program, we have striven to remediate the negative connotation of what "service" can be. As we see it, the two service courses and the minor we offer serve the needs of our department and the many departments who understand the practical value of communicating effectively. These courses also serve our university by furthering the primary mission described within my university's motto of "*Ut Prosim*" or "That I May Serve." Finally, we serve our department by demonstrating that the production of knowledge is not separate from the rhetorical acts involved in such production. Thus, even though we serve, the service we do can be and is often seen as essential, which is in opposition to those who argue that service is menial.

By embracing service as a pedagogical goal, by focusing on the scholarship of engagement, by linking theory and practice to teach students and achieve a key strategic goal of the university, we built our status and achieved recognition. Students learned to solve problems and think critically, which are not narrow, utilitarian goals; they began to realize that what they were learning had vocational, academic, and moral/social components (Boyte). One student in one of my grant writing classes said:

> English 4804 was more than just a class; it was an experience. It was more than academic; it was humanistic. This course taught me much more than how to effectively complete all the steps in the grant writing process. Each class, I learned something more about "taking life personally, letting the lives that touch [mine] touch [me]" (Remen). This course also forced me to do a little self-assessing, to look at myself and ask "how good a person am I" (Mills).

While working with my service-learning partner, Craig County Public Schools (CCPS), I developed a relationship with Mr. Stephen Janoschka, the agricultural education teacher in the high school, and Mr. Jimmy Henderson, the agricultural education teacher in the middle school. This partnership really was a "relationship between equals" (Remen).

Another said,

> If there's anything I learned the most about, it is about service and giving back to the community. I have never really volunteered much in the past, nor have I involved myself in any community before; however, when I heard those speakers and how they devote their lives to serve others within the community, I began to wonder "why?" I never really had a good answer in my head until just now. In Rachel Remen's essay (2002), she describes how lonely people begin to feel as they become older and how this quest for independence has left many unable to ask for assistance; asking for help is a sign of weakness. But she says something that I never considered: "A helping relationship may incur a sense of debt, but service, like healing, is mutual." Humans are social beings; they need each other. When you do something for someone else, you're helping yourself as well as the community. When you sacrifice, you are actually getting more back than you are giving.

Service-learning has been one very effective way our program puts its "money where its mouth is" by providing a pragmatic, rhetorical, and humanistic education. In our courses we begin with the concrete skill of editing, and the more abstract skill of recognizing that what matters about forms and genres is not "substance or the form of discourse but ... the action it is used to accomplish" (Miller 151). Following the two required courses, we offer client or service-learning projects in nearly every elective. By doing so, students develop skills and insights by focusing on complex problems that often have social and/or cultural elements.

We believe that all students need to become familiar with technologies associated with information design to authentically contribute to the community. To this end, we design assignments with pragmatic and social goals (e.g., writing grants or designing promotional materials), and we help students develop their skills as professional writers while cultivating a sense of civic idealism. Service-learning projects enable students to join with others and put their knowledge to work in the communities in which they live. Through technologically mediated writing, students gain opportunities to move back and forth between the campus and the community. Rather than making their "clients" more abstract, students are more deeply connected to them when they are able to fulfill their needs by making documents have strong visual appeal and public currency. Service-learning projects enhance students' curiosity about the people they meet and the problems they encounter.

In another sense, our current capstone course in the Professional Writing Option—Issues in Professional and Public Discourse—is also a service-learning course. It enables students to apply the analytic skills they learned while studying literature to documents in the public forum (e.g., *Presidential Report on the Space Shuttle Challenger* and *A Silent Spring*). This course gives students the opportunity to see workplace applications of their critical thinking skills and give them insight into the functions that documents (and thus writers) play in the shaping of policy.

Our curriculum is where our theories are enacted. By offering a range of courses and concluding with one that has, as its core, the goal of teaching analysis, we were able to bridge a gap between literature and professional writing; by focusing on the genre of nonfiction and its characteristics, we bridged a gap between creative and professional writing. We help English majors, our students, see the breadth of the discipline while applying what they know and learning to see from different perspectives.

As I have stated elsewhere ("Status"), our program takes a tacit tradition linked to the pejorative term of "service" and brings it into the open for examination and discussion. We teach problem-solving through service-learning, client projects, and rhetorical analysis of social and public policy. We offer a rhetorical education that has larger purposes, demonstrating that the production of knowledge is not separate from the rhetorical acts involved in that production. We value service and demonstrate its value to the university. We do not hide our relationship with service nor do we deny the value of teaching students to become reflective practitioners not only by understanding how to do essential tasks associated with writing/designing but also by understanding how the work they do and the situations contributing to that work contribute to the effectiveness of the organizations they're part of and the larger social system (28-29).

CONCLUSION

I began this essay by referring to two very different concepts: "magical thinking" and "reverence." In reality, "magical thinking" was closer to a rhetorical process involving imagination, collaboration, and deliberation. The work that emerged from that process led to a form of reverence among the many pedagogical and theoretical positions represented in our department insofar as very different positions about what English Studies is were bridged by finding common ground about "common practices." We built our program the old-fashioned way: we made arguments, relying on logic by extracting data from the university's own documents, on ethos by forging links with other departments, and on pathos by developing an inclusive philosophy and curriculum that integrated the technical and humanistic. An essential component to this work and its success was the notion that professional writing has, as all humanistic disciplines do, a larger purpose that focuses on power, people, and values or what others frame as "political, economic, and ideological tensions" (Longo 8). We discovered service-learning was a rhetorical strategy for gaining the university's heart, which became central to our understanding of the structure for our program. It provided a means of building relationships through teaching and learning, which inculcated respect. As Woodruff says, "to understand respect in a given culture, you need to look closely at how groups work together" (200). Through service-learning, teachers and students and their partners recognize that "they belong together in a common effort—to understand something that is important to understand" (202), and this "something," like Frost's "something that doesn't love a wall," has everything to do with community, with bridging the gap between theory and practice.

For our students, the theoretical becomes practical because it is related to living and working. But implementing this pedagogy isn't easy; finding the balance between service and learning is as difficult as finding the balance between theory and practice or workplace and academe. Our hope, at least my hope, is that this story will provide a text for teachers in our field, who will, after reading it, contribute to the conversation about the roles we have to teach with both pragmatic and social outcomes in mind. Service will, as I've said elsewhere ("Status"), become a concept that we can talk about, define, develop, and defend to argue effectively for our place in the academy. Such a discussion may help us see teaching as a technê, as a kind of activity in the Aristotelian sense, which has an outcome and an end or purpose (*telos*). With this conversation, we may more clearly have a conception not only of how to teach but also of why. Knowing why and helping our students understand that rationale "turns out to be a form of influence; it lies not so much in one's own operation as in the cooperation of others" (Dunne 359). A powerful result of this conversation, while perhaps not magical and surely not concluded in a year's time, will be that prospective teachers will see that reflexivity is not individual, but collaborative and that what may occur in one course, while not necessarily reproducible, will potentially lead to ideas/changes in pedagogy in other courses and in curricula as a whole. Knowing this and being able to discuss it may lead to "open-mindedness," as well as an appreciation of and reverence for what we and our colleagues do.

NOTES

[1] In 1998, the department offered, on average, only two sections of English 3774 per semester and fourteen of English 3764. Nearly all of these sections (over eighty-five percent) were taught by instructors, most with MA degrees in literature.

[2] The actual tasks, as outlined in a Request for Targeted Allocation submitted by the English department, were to
- develop and offer graduate courses in Professional and Technical Writing and Communication and in the pedagogy of these areas
- develop the writing and communication abilities of Virginia Tech undergraduates by developing and offering undergraduate courses in Professional and Technical Writing and Communication
- develop the faculty capability to offer significant Outreach activities and services in Professional and Technical Writing and Communication (1997).

[3] In year two, I proposed a special topics course (entitled The Rhetoric and Disaster and Discovery), which I taught in year three. In years four and five, we negotiated the revised and expanded English curriculum at the undergraduate curriculum committee then the department level. In year five the department approved the revised curriculum, and in year six, it was officially part of the university catalog.

[4] See http://wiz.cath.vt.edu/tw/

[5] The *Atlanta Constitution* recently ran an article listing Technical Communication in the top five fields.

[6] "the university has been slow to recognize the legitimacy of courses that emphasize the professional preparation of students" (Myra Gordon 7).

WORKS CITED

Applebee, Arthur N. *Curriculum as Conversation*. Chicago: U of Chicago P, 1996.
Beale, Walter. *A Pragmatic Theory of Rhetoric*. Carbondale, IL: Southern Illinois UP, 1987.
Bellah, Robert, et al. *Habits of the Heart*. Berkeley, CA: U of California P, 1985.
Boyer, Ernest. "The Scholarship of Engagement." *Journal of Public Service and Outreach* 1 (1996): 11–20.

Boyer, Ernest. *Scholarship Reconsidered.* Princeton, NJ: Carnegie Foundation for the Advancement of Teaching, 1990.

Boyer, Ernest. *Higher Learning in the Nation's Service.* Washington, DC: Carnegie Foundation for the Advancement of Teaching, 1981.

Boyte, Harry. "What is citizenship education?" *Rethinking Tradition: Integrating Service with Academic Study on College Campuses.* Denver: Education Commission of the States, 1993. 63-73.

Brubacher, John S., and Willis Rudy. *Higher Education in Transition: A History of American Colleges and Universities.* New Brunswick, NJ: Transaction Publishers, 1997.

Cicero: *On the Ideal Orator.* Trans. James M. May and Jakob Wisse. New York: Oxford UP, 2001.

de Romilly, Jacqueline. *Magic and Rhetoric in Ancient Greece.* Cambridge, MA: Harvard UP, 1975.

Dewey, John. "Education and the Social Order." *John Dewey: The Later Works.* Vol. 9. Ed. J. A. Boydston. Carbondale, IL: Southern Illinois UP, 1934a/1986. 175-185.

Dewey, John. "The Supreme Intellectual Obligation." *John Dewey: The Later Works.* Vol. 9. Carbondale: Southern Illinois UP, 1934b/1986. 96-101.

Dubinsky, James M. "Becoming User-Centered, Reflectives Practitioners." *Teaching Technical Communication: Critical Issues for the Classroom.* Boston: Bedford/St. Martin's, 2004. 1-14.

Dubinsky, James M. "The Status of Service in Learning." *Innovative Approaches to the Teaching of Technical Communication.* Ed. Tracy Bridgeford, Karla Kitalong, and Dickie Selfe. Logan, UT: Utah State University Press, 2004: 15-30.

Dubinsky, James M. "Service-Learning as a Path to Virtue: The Ideal Orator in Professional Communication." *Michigan Journal of Community Service Learning* 8, no. 2 (2002): 61–74.

Dubinsky, James M., and Tracy Bridgeford. "The Place of Communication in Technical Writing Programs." CPTSC 2001 Proceedings. Council for Programs in Technical and Scientific Communication, Pittsburgh, 2001: 72–73.

Dunne, Joseph. *Back to the Rough Ground: Phronesis and Techne in Modern Philosophy and in Aristotle.* Notre Dame: U of Notre Dame P, 1993.

Gordon, Myra. "Annual Report of the College of Arts and Sciences." Virginia Tech, Blacksburg, 2000.

Longo, Bernadette. *Spurious Coin: History of Science, Management, and Technical Writing.* Albany, NY: SUNY P, 2000.

Miller, Carolyn R. "Genre as Social Action." *Quarterly Journal of Speech* 70 (May 1984): 151-167.

Murphy, James J. "Rhetorical History as a Guide to the Salvation of American Reading and Writing: A Plea for Curricular Courage." *The Rhetorical Tradition and Modern Writing*. Ed. James J. Murphy. New York: MLA, 1982. 3-12.

Norstedt, Johann, and Joyce Smoot. *Request for Targeted Allocation*. Report for the Dean of the College of Arts and Sciences, Virginia Tech, 1997.

Remen, Rachel. "Belonging." *My Grandfather's Blessing*. New York: Riverhead Books, 2002. 197-201.

Rudolph, Frederick. *Curriculum: A History of the American Undergraduate Course of Study since 1636*. Prepared for the Carnegie Council on Policy Studies in Higher Education. San Francisco: Jossey-Bass, 1992.

Thelin, John R. *A History of American Higher Education*. Baltimore: Johns Hopkins UP, 2004.

Woodruff, Paul. *Reverence: Renewing a Forgotten Virtue*. New York: Oxford UP, 2001.

16 Models of Professional Writing / Technical Writing Administration: Reflections of a Serial Administrator at Syracuse University

Carol Lipson

Over a thirty-year career at Syracuse University, I have been involved in setting up programs in Professional and Technical Writing (PTW) more than once, and I have also had some involvement in helping lay the groundwork for two other programs. The contexts for these various experiences differed greatly, and in all cases local circumstances and negotiation of immediate local and surrounding campus cultures had a lot to do with the outcome of such efforts. My reflections in the pages that follow attempt to explain through example the complex ways that programs are based on human networks, not on theory and scholarship alone. I try to provide a sense of the decisions I made as I determined how best to function within the different institutional settings.

The programs I've helped develop and worked within illustrate two major frameworks. The first involves developing a PTW program within an institutional culture whose leadership structure encourages the separation and independence of program/course leaders. The individual responsibilities of such leaders are segregated (in my case, separate program leaders in technical writing, composition, and English as a Second Language). The intellectual, pedagogical and curricular agendas are developed independently, and they affect one another only tangentially. The second major framework is one in which the various strands of writing—composition/rhetoric and technical writing, for instance—are intertwined. The responsibility for leadership of each program is more distributed, less hierarchal. While neither of these approaches is inherent in a particular department structure, the first is more common in English departments, where no strong tradition exists of collaboration in scholarship, teaching, or administration. And though the second framework is more common to independent writing programs, my experience makes clear that writing program leaders can assume power in multiple ways, involving totally different degrees of collaboration with and involvement of others. Both frameworks provide opportunities and both involve difficulties.

My goal here is not to recommend one over the other, since such advice would be superfluous (faculty do not generally get to choose the history and context of their workplace). Instead, I wish to describe the experiences and concerns that arose in each situation. At the cusp of retirement, I hope this analysis is useful for those who are interested either in the recent history of our discipline's development, or for those who are facing the challenges of constructing or managing programs in similar institutional situations.

MODEL #1: SEPARATE DOMAINS, AND RETOOLING FOR PTW

I began my academic work in technical writing as a three-quarter-time assistant professor in English, soon to be hired on to the tenure track in 1979. A new Dean had become concerned that large undergraduate writing courses were being led by part-time faculty. In the case of PTW, he saw that these courses were becoming popular with different constituencies on campus, and the demand was growing beyond my ability to teach them all. Rather than ask a part-time faculty member, even one with a PhD, to train and mentor other PTW teachers and to teach graduate courses, he approved a national search for a single tenure-track faculty position.

My hiring for this position was not automatic. I had no scholarly record in technical writing, nor did I have a degree in the field.[1] That was a quite common situation for technical-writing faculty in those days, since there were few opportunities to get academic training. I don't think I was the search committee's first choice; the main competition was someone with a little more relevant scholarly preparation than I had. But to my pleasure, I did get the position. As the hiring committee requested, I soon developed an introductory graduate course required for those graduate-student and part-time teachers we were assigning to teach the 400-level technical-writing course. Because the 400-level course kept growing especially quickly, not all of the potential teachers could take the graduate-level training course in time; for these individuals, I recommended summer institutes for teachers offered on campuses such as nearby RPI, and obtained funding to support their attendance.

Quite soon, I was supervising and mentoring twelve PTW teachers, a small sub-community in an English department that was otherwise devoted mostly to the study of literature, to the teaching and practice of creative writing, and increasingly to the study of continental theory. The content of that early technical-writing course was quite common to the field, introducing students to various types of technical documents: instructions, reports, proposals, memos, etc. Teachers— mostly part-time faculty but also graduate students—had freedom to develop their own syllabi, course activities, and assignments as long as they addressed the required formal elements. From this kernel, the technical-writing teachers began to grow into a community. I soon took advantage of our successes to benefit these teachers. Since the Engineering and Management colleges were willing to pay much more than did Arts and Sciences to hire our teachers for their summer and subsidiary programs, I was able to successfully negotiate salary increases for the part-time technical-writing instructors. As a result, the technical-writing course was able to attract very strong, flexible teachers. I held a small number of group meetings, but since my own teaching load involved three courses one semester, and two the next, in a department with heavy research expectations, and since I needed to retool for technical-writing scholarship, I had to step back and concentrate on my own publication output, leaving the talented teachers to do what they did best. Because I had been on fellowship for all but two quarters of my graduate study, I was in fact more of a neophyte teacher than the high-quality part-timers I was supervising. Their expertise in teaching technical writing developed organically, as they shared syllabi and assignments. We continued to see steadily increasing demand for the 400-level PTW course, as well as increased stature on campus.

Soon I was able to establish a technical-writing advanced workshop at the undergraduate level and a related one at the graduate level; these workshops were designed for English majors, MA students, and doctoral students who might want to consider technical writing as a career. Before long, graduate students from other disciplines—all facing grim academic job markets—discovered the value of the advanced workshops. I made contacts with local industries and nonprofits to solicit projects and set up co-op positions, and a fair number of technical writers and editors began to emerge from Syracuse's English Department. Though we had no undergraduate major, graduate certificate, or graduate degree, by 1985 the PTW program at Syracuse was making a significant impact in producing technical writers.

Much of this discussion has been dominated by the first person singular. As a new faculty member, I was committed to developing my own scholarship as well as to sustaining the quality of our growing program. There was little time for collaboration. Although a composition faculty member was hired simultaneously with me to lead an advanced-composition undergraduate course and to teach graduate courses in composition, the administrative structure in the English Department tended to favor segregation of the writing programs. We crossed paths infrequently. He too was busy, arriving with three years of tenure credit – non-negotiable then. Our responsibilities were separate, and while we socialized in the first two years, he quickly faced a tenure crunch. With no senior faculty having scholarly interest or expertise in either of our fields, we received no feedback on scholarship or program leadership, no mentorship, and no advice. There was no publication such as *Design Discourse* to help us at the time. It was clear that our colleagues were attempting to determine what kinds of entities we really were, and if we were suitable as members of an elite English Department. The first department feedback we received occurred after the third-year review in my case, and at the end of the second year in my colleague's case.

In fact, an English as a Second Language (ESL) junior faculty member was hired in the same year as were my composition colleague and I. She did have two senior ESL faculty in the department, though by this period, neither of these were active in scholarly publication. She was not hired to lead the ESL program, though I presume there were expectations that she would invigorate it. Given that the ESL program leadership predated the new hires in technical-writing and composition, it's hardly surprising that the ESL program remained entirely independent from the two others, with no interaction among those in charge.

The Syracuse University English Department by the late seventies was already heavily invested in continental theory, having been influenced by the 1966 Dartmouth conference and its aftermath.[2] All new "literature" faculty positions were offered to scholars with interests and expertise in theory, even if their areas of research were located within traditional literary periods. And there were many such hires, to replace the dependence on part-time faculty for upper-division teaching in literature courses. The Dean who authorized such hiring could not have anticipated how deeply these new positions would change the face of the department. As new theoretical and philosophical orientations were articulated, traditionalist literature faculty were marginalized. It was a period of deep discord and difficulty. Yet both groups—the traditional literary scholars and the now-dominant theoretical scholars—seemed only to have a faint curiosity about what the composition and technical-writing hires were doing, with not enough investment or information to either support or confront us directly at first.

As graduate students and part-time faculty signed on to take graduate courses in composition or technical writing, discovering new approaches to teaching writing, they became extremely dissatisfied with the Freshman English curriculum they were teaching; it had been designed in the sixties and was led by a faculty member in romantic literature.[3] The two-semester course had been designed according to the best educational principles of the sixties, but was quite out of date, reductive, and ossified by the late sixties. The administrator of freshman composition did not follow the scholarship in writing studies and had no interest in changing the first-year courses to incorporate new theories and practices. But change was in the air. Other faculty in the English Department began to hear that graduate students were excited and stimulated by our graduate-level course offerings and by the curricula we developed for advanced composition and technical writing. Discourse in the English department was permeated by sophisticated new literary and cultural theories; a serious desire grew among the most powerful people in the department to change the way writing was taught at our college and to challenge the simplistic freshman English curriculum.

It is a bitter irony that as departmental attention turned to the development of undergraduate writing, my well-informed peer in composition was denied tenure. Some six months before his tenure decision he learned that his publications were not appropriate—not sufficiently scholarly, rigorous, or theoretically informed. No one had reviewed his scholarship or other work at the end of his first year (or mine), despite the fact that he had such tenure-clock time pressure. Once receiving the information, he quite understandably immediately began to immerse himself in developing two strong historical articles. As you might imagine, he was unwilling to rock any boats regarding the deplorable Freshman English situation. Many of the strong English faculty were deeply unhappy with the first-year courses and in turn disappointed in my colleague's unwillingness to challenge the issue. Institutional forces for change ultimately contributed to forcing him out of the department altogether.

On the other hand, I did not have the immediate tenure pressure, though I did have to completely retool for my new scholarly area and for my graduate and undergraduate teaching. As the child of life-long activists, I have never been good at keeping my mouth shut when I see a serious problem. Being willing to speak up, to attend meetings and participate in the push for change at the first-year level, I soon found myself elected to the department's Executive Committee, appointed to chair the Curriculum Committee, and then asked to become Director of Undergraduate Studies. In all roles, I had to take some tough positions and confront some problematic senior faculty—including the Director of Freshman English. These risks seemed to help establish my role in the department, boosting a somewhat shaky scholarly record to gain recommendations for tenure. Subsequently, as part of the department's Executive Committee, I helped develop materials for a proposal to the Dean to conduct an outside review of the Freshman English program. It took some uncomfortable years until the Dean initiated such a review, largely influenced by graduate students going public to local and campus media with their dissatisfaction. By then, I had chaired a search committee—still as a junior faculty member—to hire a strong, courageous new faculty member in composition who gave every signal of being able to hold her own in this contentious environment, and even more so of taking a leadership role to bring change. She fulfilled these expectations beautifully.

It is important to note, however, that even with the new hire, technical writing and the advanced composition courses/programs remained entirely separate, independent operations. Though both the new advanced-composition director and I, in different venues, actively addressed and explained developments in pedagogy and curriculum in our fields and made clear the deficiencies of the first-year writing courses, our efforts were separate. There was no collaboration; the model wherein faculty members operated independently held firm.

When the Dean finally approved an outside review, it included both of our programs as well as the first-year program—largely for political reasons. The review was conducted through the Writing Program Administration organization. The technical-writing program received praise, as did the advanced-composition course. A separate regular review of the English Department curriculum offered similar praise of the advanced-composition and technical-writing programs. We each had to prepare packages of materials for the reviewers, which we did independently. This isolated model of program development and leadership did lead to innovative and lively programs in technical writing and advanced composition, but at a price. The SU colleagues that I could discuss technical-writing pedagogy and curricular ideas with were part-time faculty teaching the courses, as well as graduate students. As a technical-writing faculty leader and particularly as a scholar, I remained isolated.

The outside review supported the need for change in the first-year composition program, and heavily criticized the English Department for the scant financial and other support it had made available for the teaching of writing. In response, an interdisciplinary committee, under the intellectual leadership of my composition colleague, recommended that a four-year developmentally staged set of writing courses be established, with the hiring of additional faculty in writing studies and the development of an accompanying graduate program. The English Department and college faculty supported the idea. Soon a search was established for a director of this newly envisioned Writing Program, and a director was hired to begin in fall 1986. Though the original plans involved the new Writing Program staying in the English Department, with a separate budget, in fact the new Writing Program began as an independent curricular entity with its own budget. Faculty lines were still in English, but within a few years, 60% of each faculty line was moved from English to Writing, setting up a wholly new set of tensions and opportunities for those of us associated with the Writing Program.

The ESL program did not leave English with the Writing Program. The ESL faculty member hired with me did not receive tenure, and it would have been tricky and difficult for the Writing Program to bring over the senior faculty in ESL whose curriculum was considered out of date. The English Department itself was not eager to keep this program, which was not a good fit with the now theoretically inclined English curriculum. Not long after the Writing Program became curricularly and budgetarily separate from English, ESL moved into the Department of Languages and Literatures, where it remains.

MODEL #2: SETTING UP A MORE COMMUNAL PROGRAM IN A SEPARATE WRITING UNIT, AND RETOOLING FOR RHETORIC

In 1986, after having successfully established a thriving service program in technical writing, with graduate courses and undergraduate/graduate advanced technical-writing workshops, I was once again participating in getting a new program started—not a technical-writing program, though the technical-writing service course was understood to be an upper-division 400-level component within the developmental series of four writing courses. The advanced composition course, at the 300-level, constituted the other upper-division course in the sequence. Three faculty members—the new director, my composition colleague, and I—wound up as the sole faculty in this new program, still at first technically having to teach under the old first-year course descriptions and structures, but trying out new demonstration versions of a freshman and sophomore course, designed and implemented under the leadership of the composition colleague. In the planned four-year sequence, the two three-credit parts of the first-year writing program would become a new first-year course and a newly developed sophomore course. Since virtually every student at SU had been required to take six credits of writing at the first-year level, development of the new approach to the first-year course had to be given primary attention by the three faculty members at first, while the successful advanced composition course and technical-writing course were left alone for a while. Technical writing was once again—or still—on the margins, but invigorated by the close association with the lower-division courses.

This was somewhat of a collaborative experience, with limits. The new director had been hired to develop the new courses and to create the curriculum. While she did build a sequence of courses, as the review committee suggested, she moved independently, consulting her colleagues individually but not including them in the final synthesis yielding the resulting course proposals.[4] Higher-level campus officials supported this process, though it proved somewhat frustrating to faculty, including new faculty, a few of whom were hired in the first few years before the new courses were institutionalized. The new course descriptions were very general, and groups of part-time faculty met in retreats with selected tenure-track faculty to create a variety of full versions of the first and second year courses, in order to provide a range of concrete examples and models. Besides the retreats, a prominent practice of this new Writing Program involved creating small discussion groups of part-time and graduate-student teachers, led by experienced instructors, who would meet weekly to talk over pedagogy and theory, as well as actual teaching experiences. The first and second year courses began to take fuller shape through these experiences. In addition, during some years a discussion group involving technical-writing teachers formed organically. As one of the four courses now designated as a writing studio, the technical-writing course remained popular.

In the new Writing Program environment, however, no single faculty member was attached to any one course; the courses belonged to the program, and anyone who succeeded at teaching the lower-division level could be assigned to teach the upper-division courses. People could teach technical writing with no background in technical writing or without any required graduate course as preparation. This had some obvious risks, of course. At the same time, it served to keep things fresh, as new groups of technical-writing teachers—often part-time faculty—would be assigned each semester. There would be little chance of courses ossifying in such a model, as had happened with the old first-year writing courses in English.

The demands of the new model were heavy, however. I found myself part of a small faculty that for many years carried the responsibility to administer and lead the program (I was one of two tenured faculty for the first year of the Writing Program). The amount of committee and administrative work was astounding. Notably, I was deeply involved in developing a composition curriculum and pedagogy though I was not a compositionist; this was not my area of strength or my interest. However, that's where I was required to devote enormous amounts of time and energy.[5] Inevitable conflicts developed among faculty about approaches to pedagogy and curriculum for some of the courses, often surrounding the issue of culture critique in relation to the teaching of writing, and more generally surrounding allegiances in power issues. For a number of years, the "collaborative" model was in many ways less collaborative than I and many faculty wished regarding curricular and program design decisions; and while the responsibilities were distributed, they were also much more concentrated on the shoulders of a few tenured faculty than was comfortable for a number of the faculty, and than was healthy for the scholarly careers of those with the heavy leadership-support responsibilities. I was among the latter group.

Several changes affected PTW in this collaborative model in the first few years. With so few faculty to carry out the immense needs of the program, regrettably at some point the elective technical-writing courses could no longer be sustained. At one end of the spectrum, the first two undergraduate courses were developing their identities, which took a great deal of time and energy. Simultaneously, the mission to develop a doctoral program gained force. Both of these initiatives were exciting developments and both were demanding. It soon became clear to me that at both undergraduate and graduate levels, knowledge of rhetoric would be central to my contributions as a responsible member of the faculty community. The small size of the planned doctoral program simply could not maintain a track or even a required course in technical writing, and in fact no doctoral course in technical writing has been offered since the doctoral program began in 1997. I would need to get involved in a second major scholarly retooling experience. Thus I slowly abandoned my plans to continue publishing actively in technical writing and took up rhetoric as a pedagogical and scholarly subject.[6] Technical writing was no longer as valuable to me or to the unit in this new collaborative setting, though I still kept up as much as possible with the technical-communication scholarship and published a co-edited book in the field as recently as 2005.[7] I have led and continue to lead qualifying examinations in technical communication, with two doctoral students at different stages specializing in technical communication. Despite the interest of a small group of PhD students, the field of technical communication has been viewed with disdain by many colleagues here—mostly activists for social justice who regarded the teaching of technical communication as preparing students to succeed and conform to the flawed corporate world. To say the least, technical communication was not highly valued, though this is now changing. In any case, my second retooling placed me in closer alignment with the interests of the doctoral program, as well as the needs of the undergraduate program, since the second studio until recent years was an introduction to rhetorical concepts.

The technical-writing service course was renamed Professional Writing, and its student population is now dominated by management students, focused on writing for the workplace. The engineering students are now in a minority, though we have an additional follow-up course for engineers that connects to a senior electrical and computer engineering design lab. We have also been asked to create special linked courses for Bioengineering and Chemical Engineering. And we have recently been approached to create a Business Writing course at the master's level for the School of Management. With my upcoming two years of research leave and then retirement, we would have no full-time faculty to take charge of these new developments. Thus I am most pleased that at the department's request, the current Dean has authorized a search for a faculty member with expertise in technical communication, to take place in 2008-09.

Without doubt, the design and functioning of the Writing Program at SU has from the start depended on leadership abilities of the part-time faculty, and in more recent years of advanced doctoral students assigned to administrative internships or other leadership roles. Full-time faculty are in the minority. We had three at the beginning of the Writing Program and now number eleven (two are half time with another department). The success of the Writing Program from the start depended on including the part-time faculty as partners in inventing and fleshing out courses from the deliberately brief and nonspecific course descriptions. The part-time faculty were involved in teaching the technical-writing course, whether they had technical-writing background or not; they were involved in teaching the other three studio courses, and in helping mentor other teachers. They functioned like advanced graduate students of the Writing Program in early years, and as a group they were quite concerned about loss of stature and access to faculty when the PhD program came on board.

As a result, a number of the early technical-writing courses followed textbooks to teach standard forms. Yet these same teachers would never have accepted such a formulaic approach in their composition courses. I had to swallow this, and hope that native pedagogical talent would come to the fore once the unfamiliarity of technical writing passed for these teachers. As an administrator since the mid-eighties—a period of twenty years—with just a few scattered years as a regular faculty member, facing faculty conflicts involving power and influence on curricular directions, facing serious medical issues with children, parents, and other elderly relatives for whom I was responsible, and in one period facing my own serious illness, I made a deliberate choice to just do my own teaching and not do battle on the curricular control of the technical-writing course, after one attempt at such leadership early on raised some ugly conflict-based behaviors. So technical writing—now professional writing—at SU has pretty well grown organically and collaboratively, with no particular leadership. Assignments and units that I developed were taken up by some of the part-time faculty and graduate students, but my focus involved intensive use of technology for writing early on, and this was not portable to many of the teachers. There is great variability among sections.

In 2000, a new director decided that the lower-division writing courses needed strong faculty leadership in the development of curriculum and pedagogy. As a result, there has been since 2000 a position entitled Director of Undergraduate Studies; though the title implies leadership of the upper-division technical-writing courses, the position has mainly focused on the lower division. The first director had no background in technical or professional writing, and no interest in teaching this course, nor does the recently appointed second faculty director. Only three other current faculty besides me have taught the course—one does so frequently, one will not do so voluntarily again, and one does so rarely but willingly. Some doctoral students have developed brilliant versions of the course, using it to show their abilities at course design and implementation at the upper-division level.

Yet there is not much in-house discussion about the PTW curriculum. Discussions of professional-writing pedagogy tend to be diluted amid the program-wide intensive ongoing reflection on the teaching of writing at the lower levels. This reflective discussion does feed back into a wide variety of upper-level courses such as Civic Writing, Studies in Creative Nonfiction, Style, Advanced Editing, Research and Writing, and Digital Writing, as well as Professional Writing. The strong activist bent in the department faculty has led to a small but strong strand of the Professional Writing course focused on service-learning, particularly involving community agencies. The intellectual energy of the department spent to make the lower-division writing courses more relevant, edgy, and theoretical also has helped create content-oriented courses on clusters of issues such as writing, rhetoric, and identity; writing, rhetoric, and information technology; composition, rhetoric, and literacy; as well as a course on the politics of language and writing. The future in writing is bright at this institution: a minor has been in place for three years, now involving about 45-50 students from across the campus. A proposal for a major in Writing and Rhetoric has been approved to begin in the fall of 2008, with over fifty majors signed up as of the early summer of 2009. The major includes a number of courses in PTW, as seen below, but only the large upper-division service course (Professional Writing) is offered regularly.

Since the advent of the doctoral program in Composition and Cultural Rhetoric, the tenure-track faculty have been able to, and have by now all chosen to, locate their faculty lines 100% in the Writing Program. Most of these faculty hires have been at the junior level, though a small number were brought in as tenured associate professors. As of the summer of 2008, only the original Director has full professor status, though plans are afoot for that to change in 2008-09. Yet it's clear that the heavy leadership responsibilities carried by the senior tenure-track faculty have substantially slowed their progress toward the second promotion. Especially since the faculty has grown some, we have taken considerable care not to load the junior faculty with leadership responsibilities that could impede their chances of being tenured. As the unit has evolved, we have in fact become more of a traditional department in relation to the participation of faculty. We now have a rotation of the Director of Undergraduate Studies, with leadership responsibilities for the required lower-division courses. We are fortunate to have a contingent of three former part-time faculty who serve in full-time Assistant Director or Coordinator positions. One works directly with new teaching assistants, supported by a group of 'master' teachers – chosen from the part-time and advanced doctoral-student cadres. One has primary responsibility for teacher-development programs that serve the part-time faculty and the teaching assistants, in collaboration with the Director of Undergraduate Studies. A third has primary responsibility for supporting the teachers and the program more generally in initiatives involving technology. The Writing Program Director and the Director of Undergraduate Studies meet regularly with these individuals, along with the Writing Center Coordinator, to discuss ideas and projects. Any changes in curriculum are brought before the full-time faculty, though they may originate with and be first vetted through the Directors, the Assistant Director, and the Coordinators. Committees including representation of full-time faculty, part-time faculty, relevant professional staff, and relevant doctoral students meet to develop proposals, which are brought to the full-time faculty at various stages. Except for the inclusion of part-time faculty—who have been paid to serve on committees—and professional staff, the current formulation does not differ considerably from that in many English departments. In some ways, the situation involves more collaborative participation than in the early days of the Writing Program; in other ways, collaborative activity has decreased.

MODEL #3: BUILDING A COLLABORATIVE PTW CERTIFICATE PROGRAM AND CROSS-CURRICULAR EFFORTS

What follows is probably best described as a cautionary tale. My third experience in program building for technical communication at Syracuse University resulted in an interesting curriculum, but in the end, no audience or funding for the courses and thus no implementation. This initiative began five years ago with our university's extension division, which came to us proposing that we together create a curriculum, to be delivered by the Writing Program, for an online Certificate in Technical Writing. It was meant for engineers or technical folks in industry who were assigned to handle the writing of technical documentation, as companies downsized their technical writing staffs. The faculty and administrative staff agreed to participate, and a set of four courses was developed. Four people in the Writing Program were paid to each flesh out a course to be taught at least twice each by these individuals. Funds were transferred to the department that would help us prepare part-time faculty to teach the courses. The set of courses is both attractive and suitable for the situation identified: Advanced Technical Documentation; Writing in Design and Development Environments; Information Architecture and Technical Documentation; and Technical Writing for a Global Audience. All four courses sailed through the relevant curriculum committees; they're all on the books, but have never been taught. All four were proposed collaboratively and approved for online teaching in the extension division as well as for implementation as face-to-face on-campus courses. However, the individual from the extension division who initiated the project went on medical leave, and then upon his return, the extension unit faced severe budgetary constraints—a familiar, if discouraging, story. The extension division never got to the point of proposing a certificate upon completion of the four courses. And they haven't been able to pay for offering the courses online to extension students.

Other problems affect our ability to offer the courses to campus students. As a program, we are short of willing and qualified teachers for these certificate courses, especially since at the same time, faculty across campus would like to see increasing numbers of PTW courses "linked" to their departments. We are being asked more and more to become involved in cross-curricular work in PTW. We have done so thus far using part-time faculty, but the current size and backgrounds of the part-time faculty cadre cannot sustain significant growth in the area of PTW. Teaching linked courses in PTW tends to be a politically and pedagogically challenging position, and we are confronted with the difficulty of finding part-time faculty with the myriad qualifications to make this work. [See also essays in this volume by both Anne Parker and Brian Ballentine on writing and engineering programs—editor]. We were well aware from the earliest discussions of the certificate program and of the cross-curricular requests that we would likely have to hire both tenure-track faculty and carefully selected part-time faculty to participate in teaching these courses. My experience as an administrator at Syracuse University has shown that it's best here to proceed entrepreneurially in situations such as these, which involves getting something started with little funding, showing success by attracting student interest or requests from faculty in other departments, and gaining new funds or faculty positions as a result. These certificate technical-writing courses and the requested cross-curricular linked courses may indeed take on some new life, especially with the projected hire of a new faculty member in technical communication next year. Yet as much as the Writing Program officially wants to work more closely with units across the campus in expanding and enhancing writing offerings and attention to writing, reality mediates in the availability of both personnel and funds. The difficulty arises all the more when the initiatives arise from outside the Writing Program, rather than from within.

DISCUSSION OF MODELS 1, 2, AND 3

So far, I've discussed three different program-building initiatives, situated in or across very different department cultures, and handled in different ways. In the first instance, I eventually designed a curriculum that I was proposing to focus on cultural issues, seeing technical communication as embedded in professional cultures, with their associated values and practices, and in workplace cultures and subcultures—at times clashing. I proposed that the course would help students develop their ability to understand cultures and conduct rhetorical analysis as a way to become rhetorically flexible. Some of my suggestions were woven into the fabric of the PTW course description in the early days of the Writing Program, and into the curriculum of individual sections of the course. Some others were discarded as the Writing Program priorities and contexts changed.

The one unsuccessful model (#3) illustrates the difficulty of undertaking a collaboration with outside entities, when there is no strong stakeholder among the faculty with leadership responsibilities for the effort. Since 1991, the Writing Program has not been able to secure funding or faculty positions to initiate and sustain cross-curricular efforts. Before 1991, we had a faculty line but no significant funding. The availability of funding and a faculty position would not have saved the certificate project in itself, but it would have put it on stronger footing. As Director and Chair of the department during that period, I had too much on my plate already, as did the rest of the faculty. The certificate project was no one's priority among the faculty, though one part-time faculty member took great initiative in making it happen to the extent that it did.

Of the two successful experiences in program development, the collaborative program model offered the greatest learning opportunities and growth for me (and others), though it was the riskier of the two approaches, the most vulnerable, and at times the most troubled. The standard English department model, with individuals taking responsibility, perhaps in turn, and shaping curriculum and mentoring teachers, is the easiest model to implement, the most coherent and the most conducive for course consistency, but also the least rich and varied. However, it's especially difficult to sustain a growing, thriving curriculum over the long term if one person is in charge for a long period—say as the sole technical-writing faculty member. In order to create the stimulating environment that alternating faculty leadership can provide for large multi-section service technical-writing and professional-writing courses, there would need to be more than one faculty member with technical-writing expertise or interest in the department. In small departments, that is often not possible. Even if such leadership exists, the department chair or program director must be willing to turn the course over to different leadership, and doing so is not always desirable or easy.

At Syracuse, we have also seen that the diffuse model of collaborative responsibility for curriculum—under administrative supervision, of course—can lead to some stagnation overall. This has occurred at times in the large lower-division courses as well as in the professional-writing course, though certainly not with every teacher. In the last seven years, there has been strong and clear faculty leadership of the two lower-division required writing courses, driven by the creation of a position entitled Director of Undergraduate Studies, which carries responsibility for the lower-division curriculum and for the work involved in supervising and training the new TA's who teach the lower-division courses. Appointing a highly talented scholar/teacher to this position led to changes in both the lower-division learning goals and the structures created to implement the new goals. The new curriculum developed for the inexperienced TA's—including assignments, readings, and day-to-day activities—was even taken up by very experienced part-time faculty across the program, making the course overall more uniform, more rigorous, more challenging, and more engaging for students. But no structure has been created, or will be created in any near future, for leadership of the professional-writing service course, which is well regarded across campus and well subscribed. As mentioned, the few faculty with interests in teaching this course have been involved in administrative roles with little time available for additional responsibilities. In each case, the curriculum they developed for their own teaching was not readily transferable, being grounded in special interests, expertise, and skills. Though the department has in the past shown little interest in hiring in technical communication or in Professional Writing, areas outside of the doctoral program's focus, that has now changed with my announced retirement. However, since the new faculty hire will be at the assistant-professor level, it's likely that the professional-writing courses will continue under the collaborative responsibility model, with all its benefits and faults, for some time.

MODEL #4: DEVELOPING A TECHNICAL WRITING PROGRAM IN A SCIENCE ENVIRONMENT

My career in PTW did not all take place "in-house." I have also been peripherally involved in the development of a technical-writing program next door to the Syracuse University campus, at the State University of New York campus of Environmental Science and Forestry (ESF). A long-standing financial agreement between the two campuses allows ESF students to take courses at Syracuse University, with costs escalating as usage increases beyond a certain point. Beginning in the late 1970's, our technical-writing course was both immediately popular and increasingly required by different departments at ESF. By the late 1980's, ESF found it could no longer support such an expense, determining that it would be less expensive to hire a technical-writing specialist to their own faculty and create their own program.

Yet they were cautious, for before sending students to Syracuse University, they had tried such a hire. That attempt was not a success. They discovered how difficult it was to choose, evaluate, mentor, or support a faculty member in writing, given that their faculty are primarily scientists or landscape architects. So in the late eighties, I was consulted by the ESF Vice President to help them lay the groundwork for hiring and mentoring such a faculty member. Soon, a talented part-time faculty member from Syracuse University's Writing Program, one who had already taught technical writing to ESF students, was hired into a tenure-track position at ESF. I was asked to serve on his annual mentorship/assessment committees while he was untenured, and then on his tenure committee. He was authorized to hire part-time faculty to help him, since one individual couldn't handle all the teaching. Many of those hired also taught writing at SU, and at least one has since been hired full-time at SUNY ESF. Four still keep a hand in teaching at SU. A Writing Center has been developed at ESF for its students, and a range of courses have been developed beyond the initial introductory course in technical writing. Though my work at ESF has been behind the scenes, I have been able to watch the PTW program there become a unique teaching community, with its own traditions and practices.

The new ESF faculty member began with a very different model than that implemented at Syracuse University. He had developed an innovative, challenging technical-writing curriculum for his own teaching, focused on ethical and social issues involved in scientific and technical fields, especially those related to environmental studies. In this ESF model, other teachers hired to teach technical writing were asked to follow his basic curricular structure, with his help and mentorship, though with some degree of freedom. Some of the teachers who taught at both campuses, but were left to develop their own curricular designs at SU to fulfill Writing Program learning goals, commented on the difference, noting the excitement of the experience at ESF and on the amount they learned from a brilliant curricular thinker and implementer about teaching technical communication. Two senior Writing Program part-time teachers clearly favored that model over being left to their own devices at Syracuse University for teaching professional and technical writing. And the ESF faculty—all in science or landscape architecture fields—expected a higher degree of consistency in course content than is normally the case in English departments or writing programs. The local setting here helped lead to some of the particular approaches in the development of this highly successful and growing program, though others depended on the leadership proclivities of the faculty member hired. The first faculty member hired in technical writing has remained the Director since the early 1990's, and his curricular vision and practices govern the offerings involving seven separate courses, three or four of which focus on literature related to nature and the environment.

Interestingly, the tenure-track faculty in technical writing at ESF are not expected to publish in technical writing or composition, though they are expected to attend and contribute to relevant conferences. They are rewarded for publishing poetry, children's literature, and creative writing more generally.

CONCLUSION

Each of the models I've experienced and observed in the development of Technical/Professional Writing at Syracuse University and ESF has been highly adapted to its environment, arising from particular circumstances, values, and approaches in its local culture. All have involved service courses rather than PTW degree programs. As the SU Writing Program begins its major, it also anticipates expansion of the demand for technical and professional communication offerings. In addition, the proposed major requires an internship. When I had been teaching technical writing at Syracuse University, before I became graduate director and then chair, I made the effort to line up and supervise internships and co-op positions in technical writing. This was always done on top of my load, as is frequently the case for faculty in technical writing. Now as the Writing Program envisions the expanded number of students required to do internships in writing—perhaps in community settings or in technical or business settings—the faculty are beginning to consider ways to handle the internship load that doesn't add to the already heavy demands on faculty.

My experience in a separate Writing Program with a wide range of administrative and leadership roles outside of technical writing suggests to me that being embedded in a larger writing unit can bring collaborative advantages, while also adding numerous responsibilities out of one's own scholarly or teaching areas. In my second scholarly retooling, I am now happily engaged in scholarship on ancient Egyptian rhetoric, an extension of my early work on ancient technical and medical texts. This has taken me somewhat away from my focus on technical writing publication, but has also made me a better fit with the cultural-rhetoric focus of our PhD program and department generally. The periodic scholarly retoolings I have undergone have been simultaneously unsettling and labor-intensive, but also energizing and exciting. They have without doubt slowed my progress in promotion to full professor. I would expect that most new faculty in PTW, especially those in a collaborative environment such as I found myself, would have to remain open to the possibility of retooling from time to time. My sense is that a collaborative model provides less independence and focus, yet offers broader experiences and multiple rewards for those who are willing to engage with it. At times, I envied colleagues elsewhere who had the luxury of focusing solely on publication and teaching in professional and technical writing. All told, however, I feel gratified for the risky, changing, collaborative environment here and all that it has entailed.

I would be remiss, though, if I did not emphasize finally that collaborative program leadership, especially involving fields of composition and PTW, can occur within English Departments and can be absent in independent Writing Programs, depending on the leaders themselves, on the particular faculty within the units, and on the department and campus cultures. There is no inherent one-to-one correspondence with the type of location. And a collaborative environment does not mean that collaboration occurs across the board. Here at Syracuse, collaboration occurred in certain aspects and areas, and not in others, and the specifics all changed with differing circumstances and different leaders. Collaboration in administering a writing program is of necessity a nuanced activity, affected by an array of constituent factors.

NOTES

[1] The PhD was in twentieth century British Literature; I came to SU with three years of experience as a technical writer at Caltech, as well as experience in science writing for a non-specialist audience and in writing for industry as a consultant.

[2] See Joseph Harris, "After Dartmouth: Growth and Conflict in English," *College English,* October 1991, 631-646.

[3] This individual claimed to have a letter from a prior Chancellor of SU making him Director of Freshman English for life. While no one could locate a copy of such a letter in any university files, the college was understandably unwilling to remove this individual from his position, fearing legal action.

[4] There were of course reasons for this approach. For instance, during the same period, a new Chair was hired in the English Department, having made very clear his curricular vision for the department, which sought to develop a more theoretically based curriculum. He wanted to work collaboratively with faculty to develop a concrete curricular proposal, hopefully enacting his vision. He met for two years with theorists in the department, who would not agree to implement the ideas of the individual they had hired to lead the department. The

proposal that resulted from the years of meetings was not one that any one of the individuals would have favored as the most desirable approach, but it was all the group could agree to. The process was quite ugly and nasty, and the Director of the Writing Program was determined to avoid such difficulties. Yet the approach taken brought substantial anger and resentment among faculty, having a quite negative effect on the life of the Writing Program.

[5] The administrative responsibilities in the new Writing Program did bring some course release, but unfortunately the time involved in building a collaborative new program in a highly contentious environment at SU did not come anywhere near compensating for the time and energy required. Many key groups on campus attacked the premises of the Writing Program, preferring the old Freshman English focus on grammar, on the Baker-essay five-paragraph-theme model, and on new-critical approaches to reading literature as the basis for teaching writing. The Writing Program's situation was precarious for many years. Though the technical-writing courses were never under attack, my administrative responsibilities involved the entire Writing Program venture.

[6] For instance, my work in cultural rhetoric gave rise to a collection entitled *Rhetoric Before and Beyond the Greeks,* co-edited with Roberta Binkley, SUNY Press, 2004.

[7] *Technical Communication and the World Wide Web,* co-edited with Michael Day, Erlbaum Pub., 2005.

Biographical Notes

Diana L. Ashe is an Associate Professor of English at the University of North Carolina Wilmington. She serves as Coordinator of Professional Writing at UNCW while pursuing research interests in professional writing and composition theory and pedagogy, rhetorical theory, environmental and activist rhetoric, and issues in the profession. She has published articles on environmental and activist writing, academic honesty, mentoring, and professional relations.

Brian D. Ballentine is an assistant professor and the Professional Writing and Editing Coordinator for the English Department at West Virginia University. He holds degrees from John Carroll University, the University of Rochester, and Case Western Reserve University. Before joining the English department, he was a senior software engineer for Philips Medical Systems designing user-interfaces for web-based radiology applications and specializing in human computer interaction. His research interests include professional and technical communication, digital literacy and hypertext theory, intellectual property and authorship, and open source development communities. Among other projects he is currently researching and writing a textbook, *Technical Communication for Engineers*.

Kelly Belanger, Associate Professor of English and Director of the Center for Research on Writing, Rhetoric, and Public Discourse at Virginia Tech, has been a writing program administrator at Youngstown State University and the University of Wyoming. Her publications include *Second Shift: Teaching Writing to Working Adults*, coauthored with Linda Strom, and essays on critical pedagogy, writing program development, basic writing, and professional communication. Her current research focuses on public discourse about women and sports.

Julianne Couch has been an Academic Professional Lecturer at UW since 1998. She teaches courses in the Composition and Professional Writing programs, and tutors in the campus Writing Center. Currently, Julianne is the department's composition coordinator, and works with graduate teaching assistants and faculty mentors. Her recent publications include a book review of *Scenes of Visionary Enchantment* by Dayton Duncan for *Great Plains Journal*, (2005); an essay in *Ahead of Their Time: Wyoming Voices for Wilderness* edited by Broughton Coburn and Leila Bruno (2004); and the essay "My Lewis and Clark: Discovery is at the Core" for *Heritage of the Great Plains* (2004).

Anthony Di Renzo teaches classical rhetoric and professional and technical writing at Ithaca College in New York State. A former copywriter, medical writer, and corporate consultant, he created the Professional Writing concentration for his department's BA in Writing and lead first-year writing initiatives in the School of Business and the Center for Natural Sciences. He also contributes to the Journal of Technical Writing and Communication. His scholarship explores the historical relationship between creative writing and professional and technical writing. His anthology *If I Were Boss: The Early Business Stories of Sinclair Lewis* (SIUP) was a Choice Outstanding Academic Book for 1997. Future plans include piloting a web writing course and developing a Professional Writing minor.

James M. Dubinsky is an Associate Professor of English at Virginia Tech; for the past ten years, he has directed the Professional Writing Program, a program he was hired to build. A recent winner of a college award for outreach and the university's teacher scholar award, Jim's research focuses on community-university partnerships, assessment, and pedagogy. He is the author/editor of *Teaching Technical Communication*, has contributed to journals such as the *Michigan Journal of Community Service Learning*, and he edited an issue of *TCQ* on civic engagement. Jim is also vice-chair of the board for the YMCA at VT, vice-president of the Association for Business Communication, and he hosts a radio show every Friday morning between seven and nine a.m. on WUVT-FM (http://www.wuvt.vt.edu) featuring folk, folk-rock, bluegrass, and blues.

Jude Edminster, an Associate Professor at Bowling Green State University, received her BA in English from the University of South Florida in 1977 and her MA in English from USF in 1995. She received her PhD in English with a specialization in Rhetoric and Composition from USF in 2002 for which she wrote her dissertation on ETDs titled *The Diffusion Of New Media Scholarship: Power, Innovation, and Resistance in Academe*. She serves as the faculty advisor for BGSU's STC chapter. She is currently developing an online graduate certificate program in international technical communication through BGSU's Continuing Education department.

David Franke is an Associate Professor of English and teaches in the Professional Writing Program at the State University system of New York at Cortland (SUNY Cortland). He earned his PhD in Composition and Rhetoric from Syracuse University (wrt.syr.edu) in 1999. He has worked as Director of the Cortland PWR program and now directs the Seven Valleys Writing Project (www.7VWP.com), a branch site of the National Writing Project (www.NWP.com).

Gary Griswold (PhD, Claremont Graduate University, 2003) is an Associate Professor of English at California State University, Long Beach, where since 1989 he has taught all levels of writing courses, including proposal writing, manual writing, and professional editing. In 1992, he founded the Writer's Resource Lab, CSULB's writing center program, which he has directed for eighteen years. His research interests include the history of composition studies, writing centers, innovative approaches to writing instruction, and technical and professional writing. He currently serves as both Assistant Department Chair and the director of the English Department's Technical and Professional Writing Program.

Dev Hathaway was a professor at Shippensburg University, teaching English and creative writing. He was department chair for the English Department for three years, directing the student magazine *The Reflector*, while also directing the professional/technical communications minor program. In 1998, he received the Black Warrior Review's Literary Award for Fiction. Dev was the author of numerous essays and collections of short stories. He passed away in 2005.

Brent Henze is Associate Professor of English at East Carolina University, where he serves as lead faculty in the technical and professional communication program. His research on the rhetoric of science, reporting genres in ethnological science, scientific institutions, and the scientific treatment of racial difference has appeared in *Technical Communication, Technical Communication Quarterly, Rhetoric Review,* and elsewhere. He is co-author (with Wendy B. Sharer and Jack Selzer) of *1977: A Cultural Moment in Composition* (Parlor Press 2008).

Colin K. Keeney has taught in the UW English department's composition and rhetoric program since 1988. Before returning to Laramie he worked as a writer/editor for Hallmark Communications and TIME/LIFE Books in Austin and Minneapolis, and as a freelance consultant for Ursus Ink in Albuquerque.

Michael Knievel is an assistant professor of English at the University of Wyoming, where he teaches courses in composition and professional writing. His research interests include the intersections between technology and the humanities and the position of technical and professional communication programs in the larger curricular geography of English departments and English Studies.

Carla Kungl is an Associate Professor of English at Shippensburg University, where she teaches technical writing, developmental writing, and British literature and culture. Her research interests include gender and cultural studies, the Victorian era, and popular culture and fiction. She is the editor of an e-book entitled *Vampires: Myths and Metaphors of Enduring Evil* (Oxford: The

Inter-Disciplinary Press, 2004) and the author of *Creating the Fictional Female Detective: The Sleuth Heroines of British Women Writers 1890-1940* (Jefferson, NC: McFarland, 2006). Recent publications include two chapters in books: one on Starbuck in the new version of *Battlestar Galactica* and one on the fiction of Mary Elizabeth Braddon for a collection on illness and disability in gothic literature. She has served as Director of the Technical/Professional Communications Minor since 2003.

Carol Lipson is Associate Professor and immediate past chair of the Writing Program at Syracuse University. She directed the technical-writing courses at Syracuse University from 1979 until 1986, when the overarching Syracuse University Writing Program began. In 2002, she was elected as a fellow of the Association of Teachers of Technical Writing. She has published on the history and theory of technical communication, on ancient medical writing, and on ancient rhetoric more generally. With Michael Day, she co-edited a 2005 collection of essays entitled *Technical Communication and the World Wide Web* (Elbaum). With Roberta Binkley, she edited a collection entitled *Rhetoric Before and Beyond the Greeks* (SUNY Press, 2004), and a collection entitled *Ancient Non-Greek Rhetorics* (Parlor Press, 2009).

Andrew Mara is an Assistant Professor at North Dakota State University in the English department. He began teaching and researching at NDSU in 2006. Dr. Mara earned an M.A. in Literature from the Pennsylvania State University in 1996 and a PhD in Rhetoric and Writing with emphases on Professional and Technical Writing in 2003. He combined this academic experience with on-the-job expertise as a professional communicator at Sandia National Laboratories. Research interests include posthumanism, rhetoric of technology and scientific progress, university innovation, and corporate and organizational use of new media. Dr. Mara regularly teaches the Introduction to Writing Studies and Business Writing classes. In addition he also teaches courses in Invention and Innovation, Rhetorics and Poetics of New Media, and Electronic Communication.

Jim Nugent is Assistant Professor of Writing and Rhetoric at Oakland University. He holds a PhD from Michigan Tech and an MA degree in English studies and technical writing from Illinois State University. His research interests include neosophistic rhetorical theory, the teaching of technical writing, and certificate programs in technical communication. With Lori Ostergaard and Jeff Ludwig, he coedited *Transforming English Studies: New Voices in an Emerging Genre* (Parlor Press, 2009).

Anne Parker, PhD, is the Technical Communication Coordinator in the Faculty of Engineering at the University of Manitoba, Canada, and is currently an Associate Professor. She has reviewed numerous technical communication texts for various publishers, including Wiley and Oxford Press, and has also served as a reviewer for numerous journals, including *IEEE Transactions on Education*. She has been involved for many years with the Canadian Association of Teachers of Technical Writing (an affiliate of ATTW) and has served as an editor of their journal. An active researcher in technical communication, particularly as it relates to engineering education, her current research interests include collaborative projects within the context of engineering education and integrating information literacy into the classroom. In 2004, she became a Senior Member of the IEEE, a status that recognizes professional standing.

Jonathan Pitts is Associate Professor of English at Ohio Northern University, where he coordinates the Professional Writing program and teaches creative writing, cultural studies, rhetoric, and literature. He is a 2010-2011 Fulbright lecturer in Turkey.

Alex Reid is an associate professor of English at the University of Buffalo. His scholarship focuses on the relationship between writing, pedagogy, and emerging technologies. His book, *The Two Virtuals: New Media and Composition*, received honorable mention for the W. Ross Winterowd Award for best book in composition theory for 2007, and his articles can be found in journals such as *Kairos: A Journal of Rhetoric, Technology, and Pedagogy and Computers and Composition*. His award-winning blog, Digital Digs (www.alex-reid.net), addresses issues of new media, writing, and higher education.

Colleen A. Reilly is an associate professor at the University of North Carolina Wilmington. Her teaching and research focus on professional and technical writing theory and pedagogy; electronic composition and citation; and gender, sexuality, and technology. Her publications include several chapters in edited collections and in the journals *Computers and Composition* and *Innovate* related to citation analysis, writing and technology, gender and technology, and digital research and teaching practices

Wendy B. Sharer is an Associate Professor of English at East Carolina University, where she also serves as Director of Composition. She is co-editor of *Working in the Archives: Practical Research Methods for Rhetoric and Composition* (Southern Illinois 2009), author of *Vote & Voice: Women's Organizations and Political Literacy, 1915-1930* (Southern Illinois 2004), and co-editor of

Rhetorical Education in America (Alabama 2004). Her work on the rhetorical practices of post-suffrage women's organizations has also appeared in *Rhetoric Society Quarterly* and *Rhetoric Review*.

Christine Stebbins has been a technical writing instructor at UW since 1993 and has helped design and teach the two required courses for UW's professional writing minor. Since 1991 she has also worked extensively with UW's international graduate student population. She is a contributing author in *Learning Styles in the ESL/EFL Classroom,* published by Heinle and Heinle (1995). Recently, she has designed and piloted a technical writing course specifically for international students.

Janice Tovey is an Associate Professor at East Carolina University, where she has served as Director of Graduate Studies in English, Director of Composition, and Chair of the Faculty. She holds a PhD from Purdue University. She teaches in the Technical and Professional Communication area and has published articles on visual rhetoric, and document design, both print and online. Tovey served as a coordinator for the ECU Outreach Network, training and supervising graduate students to provide community organizations with grant writing assistance. Her research interests have expanded to include ethical issues in technical communication, online teaching, and graduate programs in Technical and Professional Communication. She has served as President of the Council for Programs in Technical and Scientific Communication.

www.ingramcontent.com/pod-product-compliance
Lightning Source LLC
Chambersburg PA
CBHW020330240426
43665CB00043B/204